Agricultural robotics: part of the new deal?

FIRA 2020 conclusions

With 27 agricultural robot information sheets

Roland Lenain, Julie Peyrache, Alain Savary, Gaëtan Séverac

Éditions Quæ

To quote the book
Lenain R., Peyrache J., Savary A., Séverac G., 2021.
Agricultural robotics: part of the new deal?
FIRA 2020 conclusions:
With 27 agricultural robot information sheets.
Versailles, éditions Quæ, 80 p. DOI: 10.35690/978-2-7592-3382-3

Contents

PREFACE

One of the world's oldest professions is undergoing a massive shift. The rise of agricultural robots and other advancing technologies are helping to create a new era of farming. As farmers and producers around the world look to autonomous technologies to solve their biggest problems, the industry must rise to meet their needs.

The Global Organization for Agricultural Robots (GOFAR) actively supports the agricultural robotics sector in this endeavor by developing promotional campaigns and hosting networking events. The biggest event is the annual FIRA International Forum of Agricultural Robots, powered by VARTA AG, Naïo Technologies, the Occitanie Region and Sicoval.

This professional event brings together robotics companies and relevant stakeholders, as well as producers who seek to understand how farming continues to evolve. GOFAR and FIRA continue to prove that the great growth of agricultural robotics requires education, knowledge sharing, and collaboration.

Throughout the fifth edition of the FIRA in December 2020, more than 1,500 farmers, manufacturers, advanced technology suppliers, innovators, investors, journalists and experts from 71 countries around the world gathered to ask questions, share stories and exchange ideas. There was much to discuss.

FIRA 2020 included three days of live keynotes, workshops and demonstrations from the industry's leading pioneers. The conference recordings remained available until June 2021, making FIRA 2020 the longest running agricultural conference of the year. The Ag Robolution is alive and well. New challenges, it seems, have fostered a greater interest in these smart technologies.

The COVID-19 pandemic contributed to unprecedented changes — some welcome, others disheartening. While it pressed FIRA to pivot to an exclusively virtual conference that brought together a greater community of global participants, the pandemic also exacerbated some of the labor and food equity issues agricultural robotics are designed to address.

In addition to the complications from COVID-19, conference presenters and attendees have another big-picture problem in the back of their minds: The impending global food crisis. Producers, legislators and technology companies, in particular, are feeling the pressure to find new and innovative ways to feed the masses. Society depends on their ability to successfully and sustainably feed nearly 10 billion people by 2050.

As FIRA 2020 participants each do their part to address these larger concerns, there are many smaller issues at play. These issues include questions about the legal and ethical implications of implementing driverless tractors, whether robot bashing will prevent mass adoption of agriculture technologies and how startups and investors can work together to bring farmer-friendly solutions to the marketplace.

Each participant brings a unique perspective to the conference. Producers focus on the ways agricultural robots help them to eliminate repetitive tasks, increase efficiency and minimize their reliance on migrant labor. Startups have a vested interest in developing and marketing the solutions that potential investors will support, and farmers will actually adopt.

Technology companies work tirelessly to ensure their sensors, guidance systems and data collection platforms are precise and reliable. Legislators, attorneys and policy experts dive into the details of safely introducing advancing technologies to public and private spaces.

Taken together, these perspectives make it clear that agricultural robots are a part of the new deal. As agricultural robotics companies continue to develop innovative solutions that prepare farmers and producers to tackle the challenges ahead, GOFAR and FIRA remain dedicated to supporting and empowering the industry's future success.

INTRODUCTION

The farming and agricultural robotics industries are inextricably linked. This connection is reinforced every year at the FIRA International Forum of Agricultural Robots, where, inevitably, there are new players and new problems, bigger goals and better solutions.

As the circumstances change, the innovators, disruptors, regulators and users change, too. This growth depends on everyone's ability to pivot — whether that's to a virtual conference or to a better machine learning system. Throughout these many changes, each individual is encouraged to stay connected, both to an ever-transforming industry and to one another.

This book is a journey into an era of farming that is defined by agricultural robots. It is designed to provide a nuanced look at the industry's most pressing topics, from the over-arching impact of the global food crisis to the everyday influence of semi-autonomous tractors on a family-owned farm in Le Thillay, France. The book reveals the big picture by diving into the details. It achieves this goal by taking a deep dive into the perspectives shared by FIRA presenters and panelists.

Through their varied viewpoints, readers will better understand agricultural robots and the industry overall. These are not black-and-white problems. Armed with a better under-standing of the intersections of agriculture, technology, legislation, business and robotics, readers may come to see how effective solutions are born from the gray areas in between.

The past year provided plenty of opportunities for the world to recognize the ways in which it is undeniably interconnected. Success depends on collective knowledge, innovation and a willingness to contribute to the greater good. The FIRA 2020 conference proved that participants from around the globe were willing to show up and share their insights.

In the following pages, readers will learn not only about the ways others are making a difference, but also about how they too can lend a helping hand to move the industry forward. The book is divided into five chapters, plus a final section that synthesizes the FIRA robotics demonstrations and technical presentations into 27 information sheets that describe the features, functions and specifications of each machine.

In the first, Food and Farming are at the forefront of the conversation. "Farm machinery and sustainable agriculture must evolve together," says Josef Kienzle from the Food and Agriculture Organization of the United Nations. While robots have revolutionized farming, he explains, the future depends on their ability to do more. To that end, this section looks at the various ways that robots help farmers do their jobs sustainably, increase yields and tackle larger problems like labor scarcity, environmental waste and world hunger.

The next chapter focuses on Society. In these pages, Ag industry experts Daniel Azevedo (Copa–Cogeca), Christophe Bonno (Groupement Les Mousquetaires–Intermarché), Ole Green (AGROINTELLI) and Antoine Poupart (Bioline–InVivo) discuss the concept of robot bashing versus robot loving and how the end user's perception may impact widespread adoption of new technologies. When technology companies can provide cost-effective and reliable robots to end users, they explain, the era of robots loving is likely to take hold. In the meantime, the technologies on the market are already making a difference. "It's important to start learning together today," Green says.

The third chapter of the book dives into Regulations and Technologies. Although advancing technologies are being developed and tested, the regulations remain a work in progress. This section examines the potential issues that may arise in the marketplace and what it means to implement robotics within this context. A huge gap remains between innovation and legislation. Robotics lawyer Andrea Bertolini and John Deere's Christophe Gossard discuss these legal and ethical considerations. They note the complexities that need to be addressed when implementing advancing technologies such as fully autonomous tractors.

In the fourth chapter, experts from various technology companies weigh in about the reliability of positioning sensors and guidance systems, while farmers and industry players use their expertise and experiences to guide a discussion about the possibility of farming without tractor drivers. The overall discussion remains focused on the bigger picture: "I believe that artificial intelligence and advanced technologies can be used to make people happier," says Hajar Mousannif of Cadi Ayyad University. "By actively contributing to the AI field, we will not only ensure a better future for us but also for other generations to come. We need to put our efforts toward building tech that really matters, that solves real issues, that improves lives and, most importantly, embraces inclusion and improves our humanity."

In the last chapter, readers Go to Market. This section shares perspectives from Ag technology experts dedicated to predicting and explaining the industry's upcoming trends and business development opportunities. Representatives from startups Ecorobotix and Naïo Technologies and investment firms Capagro and BASF Ventures explain what good governance and productive partnerships look like.

This section ends with a last look at what's happening in the agricultural robotics realm right now and where companies expect to be by the next FIRA conference in December 2021. Experts from IdTechEx, Better Food Ventures & The Mixing Bowl, Sony CSL, the Yield Lab, Wageningen University and Research (WUR), Kubuta Holding B.V.—Innovation Center and Raven Industries met to share their insights. Collectively, the panelists focused on one goal: Moving beyond shiny equipment to quickly and effectively develop commercially viable solutions that address farmers' real-life problems and grievances.

The final part of the book moves beyond the theorizing and expert opinions to showcase 27 autonomous machines. These are the agricultural robots and prototypes that technology companies are currently bringing to market and that were presented at FIRA 2020.

Together, these sections provide a holistic view of the Ag Robolution in 2020 and beyond. There will always be setbacks and unforeseen obstacles, but as the conference speakers and participants demonstrate, there will always be innovation and solutions as well. This book celebrates the ways in which these ever-present challenges prompt collaboration from the world's greatest minds and markers, ultimately advancing everyone toward a more equitable future.

1. Agricultural Robots Help Farmers Feed the World

Ever since the beginning of the agricultural revolution around the turn of the 18th Century, there have always been farmers and producers that are wary of technological innovation. These "traditionalists" often have a number of reasons for clinging to tried-and-true methodologies and tools. Some resist change, others fear the unknown or, sometimes, are befuddled by the machines themselves.

This is changing. Today, agriculture technologies advance so quickly that those who resist progress are often left behind. Lightning-fast innovation has its benefits. At a minimum, the majority of farmers have embraced the idea of agriculture robots that can weed, spray, harvest and otherwise manage their least-attractive seasonal tasks. Regardless of their personal feelings, however, advancing technologies are here to stay.

In the "How Do Agriculture Robots Impact the New Deal Economy and Social Issues?" keynote, Josef Kienzle, Sustainable Mechanization Lead for Food and Agriculture Organization (FAO) of the United Nations, and Guy Waksman, Member of the French Academy of Agriculture, presented their insights.

This section details the current state of the industry, how small-scale farmers around the world are impacted by autonomous and semi-autonomous machines, and the considerations that must be made to increase robot adoption and positively impact the global food crisis.

▌ Farm Machinery Must Support Sustainable Agriculture

Kienzle kicked off the discussion on a positive note. Mechanization has come a long way, he explained, but the last 15 years have brought about dramatic improvements, including optimized design, improved digital data management and more. These improvements have also lowered costs, giving small-scale farmers increased access to autonomous and semi-autonomous technologies.

"Farm machines have revolutionized agriculture and reduced drudgery of millions of farm families and workers, but the machinery of tomorrow will have to do more than that," Kienzle says. "It will also have to contribute to agriculture that is environmentally sustainable. Farm machinery and sustainable agriculture must evolve together."

This is a key point. The global food crisis not only requires farms to produce enough food to feed more than 10 billion people by 2050, but it also means increasing food production by approximately 40 percent compared to 2012. Family farmers already supply 80 percent of the world's food.

▌ The Impacts of the Global Food Crisis

The task of producing more food to meet the growing demand must be accomplished sustainably, with farmers considering how best to manage scarce resources. The result of failing to sustainably maximize food production is two-fold: If the status quo continues, there will be more than 840 million hungry people by 2030 and fewer available resources with which to address the problem.

"The COVID-19 pandemic is intensifying the vulnerabilities and inadequacies of global food systems," Kienzle says. "COVID-19 has added additional pressure and poses a serious threat to food security. It is estimated that as many as 130 million will be added to the total number of hungry people in 2020."

Beyond having enough food, Kienzle explains, the food must also be healthy. Producing affordable, nutritious food is already a major problem. Today, more than 2 billion people cannot sufficiently or consistently access safe, nutritious food, and 3 billion people cannot afford the cost of a healthy diet. Some parts of the world experience these impacts more than others.

"If recent trends persist, the distribution of hunger in the world would change substantially, making Africa the region with the highest number of undernourished people," Kienzle says. "Innovation creates global solutions."

When there is collaboration between the public and private sectors, entrepreneurs and civil society, he adds, there is an opportunity to create the best possible solutions for the world's biggest challenges.

Figure 1: Average percentage of the population who could not afford a healthy diet in 2017 (FAO, IFAD, UNICEF, WFP and WHO. 2020. The State of Food Security and Nutrition in the World 2020. Transforming food systems for affordable healthy diets. Rome, FAO. https://doi.org/10.4060/ca9692en), Joseph Kienzle, FAO.

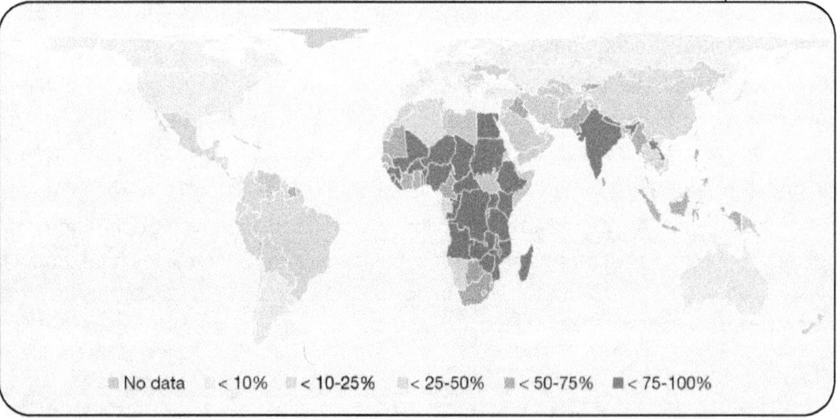

▌ Holistic Solutions are the Product of Innovation and Education

Innovation and agricultural robots can help address these problems in a number of ways. Not only do they contribute to sustainable agriculture practices, Kienzle says, but they also support and empower family farmers and other small-scale operations to grow food productively and profitably.

"Innovation is key to feed the growing population and face the challenges of climate change," he adds. "It can be instrumental for early and timely planting, especially when there are labor shortages, and it can help prevent food loss when harvesting.

"Due to climate change effects, the planting season is shorter, and the dry spells are more frequent today," Kienzle continues, "we have scarce natural resources, and through the use of appropriate equipment, innovation and robotics can contribute to the precise use of resources, in particular, inputs such as fertilizer, seeds and pesticides, in this way avoiding overuse and waste."

In order for agricultural robots and other advancing technologies to make a difference, they need to be adopted and used in farms around the world. FAO is helping to do this now. The organization's Hand-in-Hand is an "evidence-based, country-led and country-owned initiative to accelerate agricultural transformation and sustainable rural development to eradicate poverty and end hunger and all forms of malnutrition."

According to the Hand-in-Hand website, "the initiative prioritizes countries where national capacities and international support are the most limited or where operational challenges, including natural- or man-made crises, are the greatest."

Without buy-in from the country affected by these issues, the initiative would likely fail. Hand-in-Hand's cooperative approach prioritizes ownership and leadership to unlock agribusiness potential. As Kienzle notes, it is also helpful for bridging the rural divide and giving women and youth increased access to markets and information technologies.

Conservation agriculture practices can be introduced in other parts of the world as well. Kienzle draws attention to dust bowls that have been created through wind erosion in countries like Turkey, Russia, the United States and Germany. When fertile topsoil is pulverized as a result of heavy tillage, it also becomes more prone to water erosion. The water sweeps the soil into the sea, while creating gullies on land. In turn, this upsets the region's ability to farm.

FAO helps to remedy these issues by teaching and promoting sustainable farming knowledge and practices in these regions and through other initiatives around the world.

"In the oceans, it takes more than a thousand years to build one millimeter of fertile topsoil," Kienzle says. "Soils are, therefore, considered a non-renewable resource. Farm machinery needs to be lean precise and efficient in order to minimize the impact on the soil and landscape."

The right tractor can prevent topsoil pulverization. These machines tend to be lighter and less expensive, which makes them easier for small-scale farms to invest in them. Meanwhile, using targeted applications of pesticides or eliminating their use entirely can preserve soil moisture and limit soil disturbance.

"FAO aims to promote agricultural knowledge and practices and improve the sustainability and resilience of farming systems of smallholder farmers through ecosystem-based approaches," Kienzle says, noting that this work cannot be accomplished alone.

"In order to scale up and to allow for the uptake of technologies and sustainable practices, it will be important for all actors and stakeholders to work together," he continues. "This includes defining their roles. Of key importance is the role of the private sector and the role of the government which is developing policies and regulations to create an enabling environment for technologies to flourish and reach those who need it."

▊ Global Problems Require a Collaborative Approach

Advancing agricultural knowledge in this way is one example of combining efforts to move toward greater sustainability. Similarly, Kienzle urges the industry's many players to collaborate on viable solutions that address other conservation and farming issues, too. One way to help is to educate and support small-scale farms trying to adopt mechanization and get their businesses off the ground.

Kienzle understand that farming must be a profitable endeavor. Enhanced market access helps farmers in places like sub-Saharan Africa get the machines they need to succeed with sustainable crop production, but this too requires a collaborative approach. FAO has combined efforts with the African Union Commission (AUC) to create a framework that "points the way towards addressing challenges and creating new opportunities to ensure the successful adoption of mechanization."

This effort directly contradicts a common worry within the agricultural industry: Robots will replace human labor. Kienzle says this fear is unfounded. Instead, robots create new and better jobs along the value supply chain. Ultimately, the move toward mechanization has the potential to bring new talent into the industry. Kienzle sees the possibilities.

"While agricultural robots are still in their early stages, there are very clear indications of their potential," he says. "The challenges ahead are not only technical, but also socioeconomic, in particular, with regard to capacity building and the need to understand the principles and technologies involved."

Through their many programs, FAO is dedicated to confronting the farming, sustainability and food scarcity obstacles that threaten to become insurmountable in the years ahead. Those who choose to join this fight will have a powerful tool at their disposal.

▊ The Path to Adoption for Agricultural Robotics

In Waksman's presentation, he touched on some of the same issues Kienzle reported. Namely, that the COVID-19 crisis has complicated matters. Farmers often worry about losing their jobs, he says, but closed borders created limited access to markets and a worker scarcity that caused the opposite problem. In the future, machines can be employed to pick up the slack.

"Robots are always a symbol of the ultimate age of technology," Waksman says. "Between machinery and manpower, I believe that the nations with robots are able to maintain new diversity of production under acceptable economic conditions."

The next generation of agricultural robots is also being developed to address some of the adoption-related hurdles the current generation experiences. Waksman says that robots must be adapted to the regulatory content so that they may eventually be able to work alone in fields, and farmers may even need to restructure their fields in order to accommodate the machines.

He also believes that these robots require sophisticated software that will enable them to process data from cameras and sensors. There will likely be a learning curve for operators as they learn to use these increasingly complex technologies. The next generations of robots, however, can make up for certain deficiencies with creative solutions.

"Right now, robots are more the tortoise than the hare," Waksman says. "They move slower, but they are able to work longer. Some work throughout the night, which makes up for their slower pace."

▌Advancing Technology Suppliers Can Support Adoption

Picking up where Kienzle left off, Waksman discussed the ways that advancing technology suppliers can capitalize on a robot's potential. He shared his perspective about how these suppliers can work to bring their designs and prototypes to market.

Many farmers may be on board with the idea of agricultural robots, but innovators that want to attract early adopters must go above and beyond. This means further testing, demonstrations and a thorough understanding of what prevents farmers from buying in right away.

Figure 2: Robots create new and better jobs along the value supply chain, Joseph Kienzle, FAO.

"Technical progress must work together with promotion, testing, demonstration, and insertion efforts in the digital world of farms," Waksman explains. "Demonstrations help to convince farmers of the benefits of robots. They also ensure that the equipment works properly and that the sensors, drone imagery and other components are effective."

▮ Agricultural Robots Benefit Farmers and the World

Waksman's description of agricultural robots as high-technology machines aligns with the keynote's final message from French Minister of Agriculture Julien Denormandie, in which he described agriculture as a high-technology profession. Such technology, he says, "must be used to reduce dependencies and recover our agro-food sovereignty."

While Denormandie spoke primarily of France and the investments the country has made in the future of agricultural technologies—including a 15-million-euro investment in research and development, as well as the facilitation and deployment of new agriculture technologies—his words echo around the world as a call to action.

All countries rely on limited resources like water, and all countries are impacted by the negative effects of climate change. There are mechanical and technical solutions to these problems, Denormandie says.

"Big technology, big robotics are sectors of the future," he continues. "We have many examples at the forefront of this high technology. I think, for example, of the electric hoe or the weeding robot. These tools are technological jewels, involving both innovative sensors and systems based on artificial intelligence.

"As a result," he continues, "the weeding robot is now extremely precise, and selective weeding is obviously a benefit for the farmer in terms of work and facility, and at the same time, for the approved ecological transition we all wish for."

There are opportunities for everyone to benefit from investments in the agricultural robotics sector. When organizations, companies, associations and everyday people come together, these projects quickly become solutions. Denormandie underlines the importance of the industry's innovative future.

"I believe in robotics," he says, "and I think that it is a sector that we must support ever more strongly."

2. Successful Robot Adoption Depends on Reliability, Security and Trust

Autonomous technologies have a lot to offer farmers. In addition to the oft-cited benefits of reducing the reliance on migrant labor and eliminating repetitive or dangerous tasks, agricultural robots also bring new possibilities to an age-old industry.

In the "How to Go from 'Robot Bashing' to 'Robot Loving'" keynote roundtable, Daniel Azevedo (Copa–Cogeca), Christophe Bonno (Groupement Les Mousquetaires–Intermarché), Ole Green (AGROINTELLI) and Antoine Poupart (Bioline–InVivo) discussed what manufacturers, farmers and end consumers require in order to reap the full benefits of the Agrobolution.

The panelists offered a wide range of perspectives based on their areas of expertise. Collectively, the group represented farmers cooperatives, entrepreneurs, technology companies, and investment firms. Despite these differences, the group agreed on many points. Most notably, they are all devoted to taking the steps necessary to advance and support a robot-loving industry.

This section details what each sector of the industry must understand about today's agricultural climate and how everyone can work together to succeed in the long-term.

■ The Ag Industry Already Loves Technology

While it may seem surprising that an industry born from hand tools and cattle-drawn plows has embraced technology, that's exactly what's happened. From tractors and sprayers to shaking machines and mechanical harvesters, agriculture has become increasingly tech-savvy. Autonomous technologies are simply the next step.

According to Daniel Azevedo of Copa–Cogeca, a group that unites European farmers and European agri-cooperatives, robots are already a big part of the farming experience in the EU.

"Agriculture is very clearly devoted to technology," he says. "I think most people will be surprised how agriculture is using technology now."

The rise of automation in the farming sector is interesting for a number of reasons, Azevedo notes. Robots and other advancing technologies are poised to help producers deliver on the new policy framework that focuses on innovation, job creation and competitiveness in the world market. They also make it easier for farmers to provide a high level of food security, all while following sustainable practices that maximize welfare and reduce environmental waste.

An abundance of benefits means robots continue to be attractive to many farmers. Few are worried about the impact on jobs because, as Azevedo points out, "it's up to the farmer to decide which technologies he is going to use."

Robots can also be a means to move the industry forward. Farming is likely to attract new and younger talent to an increasing technology-driven profession with fewer monotonous tasks and shorter workdays. Azevedo says the latter is already happening.

"Robots organize the work of a farmer," he says, "and we have seen examples of how the robots are actually helping families to have breakfast together in the morning because the work is now more structured."

If anything, Azevedo explains, it's society that has shied away from accepting the latest generation of agricultural technologies.

"It brings up an interesting question," he says. "Why do we accept the use of technologies in all aspects of our life, but when it comes to agriculture, concerns are raised over safety? These are technologies that are being used and are considered safe and have been proven around the world."

In order to alleviate the public's concerns, Azevedo continues, robots must prove they can deliver on their promises to improve working conditions, increase profitability, and ultimately, get the job done. Similarly, the biggest key to keeping farmers onboard with the latest innovations is to ensure they actually work as designed. Nothing inhibits adoption rates more than big promises followed by second-rate solutions. If companies build it properly, Azevedo says, farmers will come around.

"I can guarantee that if the robot will deliver and do what the farmer needs, they will learn immediately how to use it," he says. "I can tell you I've seen farmers that never cared about technology with mobile devices and iPads in their hands because it delivered what they needed, and they are curious."

Robot Bashing is an Opportunity

Not every farmer, however, is meant to be an early adopter. There will always be people who prefer to employ old-school practices and traditional methodologies.

"I think that robot bashing is not really a problem; it's an opportunity," says Christophe Bonno of Groupement Les Mousquetaires–Intermarché, a group that works to consolidate production chains and bring together agricultural partners and the distribution network. "It's important to explain our robots can provide safety, confidence, transparency, and a greater possibility to address consumer trends."

"It's a new opportunity to build up confidence with the consumer," he continues. "We want to develop green agriculture by reducing the use of chemicals and the carbon footprint for crops treatment. We also want to ensure ground fertility with better water management and more biodiversity. This communication must be our number one priority."

Robots also offer a persuasive case for improving product traceability and overall quality. When these benefits are articulated to the end consumer, it builds support for what farmers are doing. The consumer-farmer relationship is an essential part of building a society that loves robots.

"The consumer trusts the farmer but not his practices," Bonno says. "We want to work on building that trust. With robots and new technologies, we will be able to match treatments to crops based on composition. We will be able to develop artificial intelligence which uses computer programs to analyze data."

Although the full range of benefits are not yet known or understood, Bonno continues, the data can be used to control product traceability and provide critical information to consumers when they need it most.

"We should be able to earn the customer's trust with all this information," he says.

▌ Farmers, End Consumers Have High Expectations

Trust is one aspect of building a robot-embracing industry. Meeting expectations is another. When it comes to satisfying farmers and their customers, technology companies have their work cut out for them.

"The use of robots is of no interest to the end consumer," says Antoine Poupart, Bioline–InVivo, an agricultural cooperative group. "The expectations are around results: What's on my plate? What's in my environment? What's happening in my society? It's not about the means of production. So, we need to ask, "which of the end consumer expectations can be met by robots?"

This disinterest in robots and their role in agricultural production, however, may be changing. Thanks to the COVID-19 pandemic, says Ole Green of AGROINTELLI, an agricultural automation company, people are beginning to pay attention. He encountered this heightened awareness recently when talking with end consumers about vegetable and berry production.

"We suddenly have a society looking much more into hygiene factors," Green says, "so, the fact that you could actually start harvesting your crops without actually having any contact or minimizing your pollution risks that could be an element where automation could really start to support the hygienic factor in our high-value crop farming."

Value is also an important part of the new agricultural revolution. Farmers are invested in securing market value, while consumers want to purchase products with maximum value. Poupart says that according to the studies and market research he's seen, there are three pillars of expectations that customers value more than others.

"The first is food quality, which is mainly around safety and nutritional quality," he explains. "The second pillar is around environmental impacts, mainly about climate change and biodiversity. The third, which is really increasing in importance, is about welfare and quality of life at work. Those are the pillars of value and customer expectations that we really need to focus on and have robots provide an answer to."

Since the customer can't always be on the farm to see robots and other technological solutions in action, the best means of providing value is through information.

"Farmers today have very few tools to provide information to the consumer," Azevedo says. "We hope that the data will not only be collected by the robots but that the technologies will actually enable us to provide further information to the consumer. It is also extremely important to get information from the consumer, so that we can use it in our farms and adapt our practices accordingly."

▌Robots Must Be Reliable, Valuable and User-Friendly

For farmers to effectively meet their customers' needs, it is essential to know their preferences, desires, and buying habits. This information also helps producers to better define what they require from their robots. Certain baseline capabilities, however, are critical for farmers to begin adding these autonomous technologies into their workflow.

"There's no doubt that when introducing a new technology, the robustness is one of the key factors that we have been met with," Green says, adding that farmers inquire about how the robots compare to a machine they've known for 40-plus years: a tractor.

"Are the robots reliable to do the work because the farmers know that the work has to be done within few days or a few hours when the conditions are good," he says. "The new technology has to be reliable before this is something that really will have a big impact on the market from an application point of view."

In addition to reliability, farmers need to know the robots can perform as well as, if not better than human laborers. While autonomous technologies can help farmers reduce the negative impacts of labor shortages, product quality matters. Operations that want to succeed long-term need to continually maximize the marketable yield.

Robots, as Green notes, have the ability to ensure uniform quality in the sellable product. When seasonal workers come and go each year, the farmer must constantly train new people to manage the same processes. If most of the labor isn't retained from year to year, the quality may suffer.

These seasonal jobs can also be replaced by better-paying, more attractive positions, such as "machinery assistant" or "robot repair specialist." Changing the job market within agriculture is an example of how robots can be used to create value for the farmer.

"If you are actually able to get some money from your profession, you will be able to employ other people and you will be able to expand your activities," says Azevedo. "So, I think the conversation needs to go in the direction of, what kind of tasks could be automatized and how can we create value to make sure that the business maintains a profit, is able to deliver, can pay taxes and is able to keep bringing jobs into rural areas? The farmer will make the necessary investment if it delivers a value."

By automating the repetitive tasks, farms can increase productivity with the same size crew. With the robots managing the grunt work, agriculture professionals have the opportunity to focus on agronomy and other things that require human expertise.

"We are much better to look at variability in the field, soil, animals or plants than spending time in training expensive computer networks," Green says. "That's where we should spend the time, in visiting the spots in the field to find out why are they not yielding like we would like them to. It is really important that we value the farmer's understanding of agronomy and business, and then it's up to us as manufacturers to make sure that the solutions are robust, reliable, and easy-to-use products."

The case for introducing robots onto the farm is a strong one from the value standpoint, but adapting to new technologies can be daunting, especially for those who have become familiar with the tools already at their disposal. That's where advanced technology suppliers need to do the prep work to simplify their machines.

"I believe that the farmer shouldn't need to be a software programmer," Green says. "If we are in that situation, then we didn't finish developing the robot interface."

The majority of farmers, he says, want one or two buttons on their machines. Engineers love technology and developing advanced tools. Farmers crave ease of use. The end solution should give operators the same feeling as getting into a new car, Green says.

"You can get into any new car and know by intuition how to use it, and I think that is what we would like to see with technology in agriculture."

▌ Tech Companies Must Build Trust Through Responsible Practices and Customer Support

Ideally, the relationship between farmers and robotics companies continues long after the sale. Just as consumers need to be able to trust the farmers that supply their food, farmers need to be able to trust the manufacturers who supply their machines.

"I think trust is one of the big challenges we have right now, and all robotics manufacturers have a huge responsibility," Green says, "I think that's really where the honesty of the new players on the market has to be true and valid. If we are over selling the technology, we will be shooting ourselves in the foot because the farmers and the consumers won't have the same experience as we are trying to sell it."

The best way to minimize buyer remorse is to ensure users understand how to operate their robots. Green equates it to selling other common agriculture technologies like autosteer or a sprayer. The onus is on the manufacturer to confirm the customer is armed with the information necessary to effectively use the equipment. Training and support are essential, but Green also views this as an opportunity for manufacturers to set themselves apart.

"There's no doubt that the competition between robotic manufacturers todays is not only the physical design or the way its constructed. It's also the usability. It's whether the user interface is easy to address by someone who didn't have before. This is, of course, also a competitive parameter for the robotic manufacturers."

This responsibility to the farmer extends beyond the hardware, too. Although this is an important component, one of the biggest concerns plaguing today's robotics operator centers on data ownership and security.

"The farmer is in a situation where he is making investments in terms of buying the technology or buying the robot, and it will be the one collecting the data," Azevedo says. "Data that is collected during farming operations or on the farm should belong to the farmer. He should be the one controlling the access to the data, and that is also to ensure that he gets some of the value that is created by sharing the data. We need to make sure that the farmer is not the only one taking risks and making the investments."

While data sharing provides value beyond a single farming operation, Azevedo adds, the farmer will need to be a part of that process. There is already a general mistrust around data usage. Green says that manufacturers have a responsibility here as well.

The farmer must have the power to decide who can use the data collected from a robot on his operation, but that data will be managed through a different system. Green compares the process to selecting a bank. Farmers will need to decide which vendor to trust with their valuables.

Another complication is the sheer amount of data a robot can collect. According to Green, one of AGROINTELLI's standard robots can produce anywhere between one half and three terabytes of data per day. Not all of what's collected will be useful.

"I think there's a big challenge in actually sorting out what is data and what is information," he says. "That's a huge amount of data and not even a 4G or 5G network will help transmit all these data. So, we really need to have a referral insight into how we make this informative for both the farmers and the end consumer."

▌ There's a Lot of Potential and a Lot More Work to Do

Robots have already made their way onto farms across the world. From autonomous weeding prototypes to fully functional feeding robots, the agricultural industry is in the midst of another technological revolution.

"I think one important thing that we've seen in this very strange year is our reliability of workforce," Green says. "There is shortage of people that can drive our machines up and down the field, and this is the core element where robots can make a difference today. A lot of what we talk about is future perspectives, but robots are on the market today, and they can actually make a difference today."

This is merely the beginning. The panelists agree that there is a lot more work to be done.

"We need investment," Poupart says. "We need investment from the EU policy framework. We need investment on broadband, so that farmers have access to and can actually share the data that is provided. We need access to internet infrastructure. We need social funds to support training. There are a lot of new policies that will actually level the playing field to make sure that the farmers can actually choose the technology that best fits their needs."

Azevedo believes the industry also needs to focus on the next generation of farmers and robotics operators. Those who miss out on early learning opportunities are likely to be left behind.

"We need to prepare the new generations for all the technologies that are available," he says. "You need to have a good basis for understanding as a kid, then you need to have

vocational training and advisory systems that provide support, not only for the robotics but for data processing and how to integrate other related activities. This is extremely important."

As impressive as the current technologies are today, they're going to get faster, smarter and more sustainable over time. This generation of robotics most helps farmers complete singular chores, but Green says that artificially intelligent systems are not far behind. The massive amount of data that's being collected now will transform task-master robots into savvy decision makers, provided the infrastructure can keep up.

"It is extremely important the access to good sources of reliable broadband," Green says, echoing Poupart's sentiments. "We want to avoid these gaps between rural areas and urban areas. Rural citizens should have access to the same services and rights as urban citizens. I've seen a number of research projects right now addressing how we can benefit from 5G. I would say if I could just have 3G all over the agricultural domain area in Europe, that would be a good starting point."

Regardless of the obstacles to mass adoption, the panelists agree this is an exciting time for the Ag industry. The future lies in the hands of the next generation. Azevedo encourages them to continue moving things forward.

"The agriculture sector is exciting," he says. "We have shown resilience during the COVID-19 crisis. We have kept the food security and kept the food coming to the plate. We never stopped working. It is important to talk to your decisionmakers, to the politicians and make sure that they invest in our farms and in sustainability, and that farmers have access to the latest high-tech technology. With access to those technologies, we will be able to deliver."

3. Farming with No Tractor Driver, is it Possible?

One of the most highly anticipated agricultural robots, driverless tractors present an interesting conundrum for everyone from advanced technology suppliers to legislators. Most industry experts agree that the technological capabilities are there. It is more a matter of "if" than "when." The problem, then, is less a matter of how to develop a driverless tractor and more about how to regulate one.

While farmers are generally eager to embrace an autonomous machine that eliminates one of their most time-consuming tasks, there is a wide range of potential issues that must be considered before such machines make their way onto the market. In the FIRA 2020 keynote, "Farming with No Tractor Driver, is it Possible?" two presenters and four panelists shared their insights.

From the legal perspective, Andrea Bertolini shared the many ways legislators might regulate autonomous tractors. Christophe Gossard from John Deere GmbH & Co. KG focused his presentation primarily on risk assessment and safety and security concerns.

In the roundtable that followed, Antoine Chatelain (nursery employee and robot operator), Aymeric Barthes (CEO of Naïo Technologies), Gordon Clements (General Manager Solutions for VARTA AG) and Greg Meyers (CIO/CDO for Syngenta) discussed the various limitations that needed to be overcome for driverless tractors to be developed, deployed and, eventually, adopted worldwide. This section looks at each of these viewpoints in greater detail.

▌ Civil Liability, Time Machines and the Case for Insurance

With the implementation of any new technology, there are unprecedented risks. In this way, driverless tractors are no different from weeding robots, robotic vacuums, or long ago, the first automobiles. Whenever a machine causes harm to an operator, there must be a legal means for compensating the victim.

"Civil liability is basically defined as the set of rules that determine who is bound to pay damages to compensate the victim when the victim suffers harm," says Andrea Bertolini, a lawyer specializing in robotics. "At European level, typically liability, and civil liability in particular, requires fault. Basically, that the party that is being held responsible can be considered negligent or violating a rule of law."

With respect to robots and advancing technologies, fault can be considered in a few different ways. The first is based on whether the machines themselves can be held responsible for causing harm. According to Bertolini, there is some debate about whether advancing technologies that can "learn" like robotics and artificial intelligence applications can be deemed agents, and therefore, held accountable for causing harm. He cites a provocative article from years past that posited even a smart thermostat that automatically adjusts the temperature in a room could be considered an agent and held liable for harm.

Bertolini is not convinced by such proposals. Because today's machines are not sophisticated enough to be agents. They have neither the skills nor the abilities to cause harm. Under the current legal framework, he says, the programmer or producer of the applications would more likely be held responsible.

The issue became slightly more complicated when in 2017, the European Parliament considered a proposal about whether to attribute "electronic personhood" to advanced machines.

"Some people understood this to be the acknowledgment that machines are agents and therefore self-responsible, but that was actually not the case," Bertolini says. "What our European Parliament was proposing, among other alternatives, was to attribute personhood like it does to corporations also to some kinds of advanced machines. That would be a purely instrumental concept."

Since that proposal, there has not been much follow-up, and advancing technologies do not have electronic personhood according to the existing legal framework. Bertolini proposes that driverless tractors and similar machines be considered in a different context: the words of Professor Paolo Grossi. The Italian law professor and Italian Constitutional Court judge famously said, "the law is not a brick that falls on our heads; the law is all around us."

"This is to say that if overnight, somebody invented the time machine to undergo time travel, that would immediately and already be regulated," Bertolini says. "Meaning, that if somebody suffered harm while using the time machine, even if there is no time-machine-specific regulation, those that got harmed using the time machine could sue the producer of the time machine in the court, and the judge would have to find a solution to that case with existing norms that are already available.

"This also allows me to say that it is not useful to debate whether or not we want technology to be regulated because technology is already regulated the moment it comes into existence," he continues. "Rather, we are only debating what kind of legislation we want for it."

Today, all machines fall under two different liability norms: product safety regulation and product liability directive. Product safety regulation, Bertolini explains, is "that body of requirements that are set through primary legislation at the European level to determine what parameters a given application needs to meet in order to be considered a safe and also to be then sold legally on the European market."

Advancing technology suppliers must comply with these standards to sell their products on the market, but these parameters do not limit the company's liability. That's where product liability directives come into play. These directives aim at two things: ensuring product safety and ensuring victims who suffer harm are compensated. When the user encounters harm, the producer of the good is strictly liable. The victim does not need to show negligence, only that the product was defective. This is tougher than it sounds.

"There are grounds to doubt that, at European level, this body of legislation is functioning properly as it as it should," Bertolini says. "There are a number of studies showing that there is a very limited litigation that allows the application of those norms. It's very hard for claimants to demonstrate that the product was actually defective and therefore that the producer is obliged to compensate, and this causes a lot of concerns when it comes to advanced technologies."

When victims have little recourse against the products that harm them, there are fewer early adopters of new technologies.

"It seems to be a useful thing in a producer's perspective to say, 'if it's harder for victims to sue us in a court, we will be favored somewhat,' but that is not actually the case," Bertolini says. "I did studies to show that this will slow the uptake of technology in many cases because if users do not feel safe around those kinds of applications, they might not want to switch to the more autonomous versions. They would rather keep the more familiar technologies."

Instead of relying on product liability directives, Bertolini suggests a more practical risk management approach. Take the example of driverless cars, as they have similar characteristics to autonomous tractors and other advancing technologies. If the producer of the technology was held "absolutely liable under all circumstances," Bertolini says, "he would acquire insurance."

The cost of that insurance would then be transferred onto the product's users who would pay for this insurance as part of the cost of the product. This would result in compensation without litigation for any potential victims. The costs get distributed along the value chain through contractual agreements.

In the case of the driverless car then, this evens the playing field between the advanced technology and a traditional car. Now, users are more likely to choose the driverless option than if there weren't a framework for addressing liability and damages.

While this is a good option for addressing harm within the current legal framework, Bertolini believes insurance has its limitations. There is more work to be done.

For one thing, there has to be a market for this specific type of insurance, as driver's or homeowner's insurance is unlikely to address the unique problems created by advancing technologies.

"I do believe that all robotic applications are different from one another in that you cannot really find a one size fits all solution like the European Parliament is attempting to do," Bertolini says.

For example, agriculture robots generally exist within segregated environments, away from the general public. The driverless tractors will not need to account for the erratic behavior of other drivers, as a driverless car would.

"This is an element that should be taken into account when considering a regulation for agricultural or robots," Bertolini says. "It's always good to take into account the specific technological features of each machine. I always have this bottom-up approach, and I do really think that it is necessary for conceiving new legislation. So, even when it comes to insurance, it is a tool to address some of the liability concerns, but it is not always the best tool."

▍Safety, Security and the Journey to Market

From the manufacturer's perspective, John Deere's Christophe Gossard presented the two types of challenges that needed to be addressed before driverless tractors become a real-world possibility. The first is managing safety and security; the second is ensuring the legal requirements to get autonomous machines onto the market are not cumbersome.

When it comes to product safety in Europe, Gossard says, risk assessment is done through the manufacturer. When there is no standard available (especially for newer technologies), there is likely an international or industry standard that can be applied. This is very different from places like the U.S. where risk assessment is conducted in court through a team of lawyers.

For new technology risk assessment, the process is supported by the original equipment manufacturer (OEM). The OEM's job, Gossard explains, is to validate the conformity of equipment that can be used in the fields and may potentially circulate on the public network.

In his presentation slides, Gossard notes that manufacturers can use European and international standards to allow proper risk assessments under the future regulation for machineries replacing the current Machinery Directive (2006/42/EC). From there, manufacturers can integrate new technologies (IoT devices, AI/ML, cyber-security, autonomous features, etc.) into the Essential Health and Safety Requirements, while maintaining high safety and security requirements, and protecting the OEM against potential litigations.

When it comes to connectivity, Gossard contends that there needs to be a robust architecture based on a Permissioned Distributed Ledger (PDL)/private blockchain. A PDL combined with a certification process will provide trust to the end user, he says, while ensuring openness to interconnect other equipment completing the eco-system required to make run an autonomous vehicle.

The new legislative framework and declaration of conformity are very important for an industry that wants to deploy autonomous vehicles, Gossard says. There are benefits, Gossard explains, including self-certification, removing market barriers for the industry and SMEs, cutting unnecessary costs and avoiding a disproportionate administrative burden for the manufacturer. He believes that it doesn't make sense for small operations to have a full legal team to deal with any issues that arise from advancing technologies.

As the industry gets closer to bringing driverless tractors onto the market, new challenges will arise. One of the major issues around advancing technologies centers on data. The sovereignty of the data is really at the core of the discussion right now for the European Commission, Gossard says, and there is a lot of discussion from Germany and France about the architecture of standards for developing new products and new technology, especially the data created at the sensor level.

No one knows what will come of these discussions. Gossard, however, prefers to focus on the future, dealing with the issues as they arise.

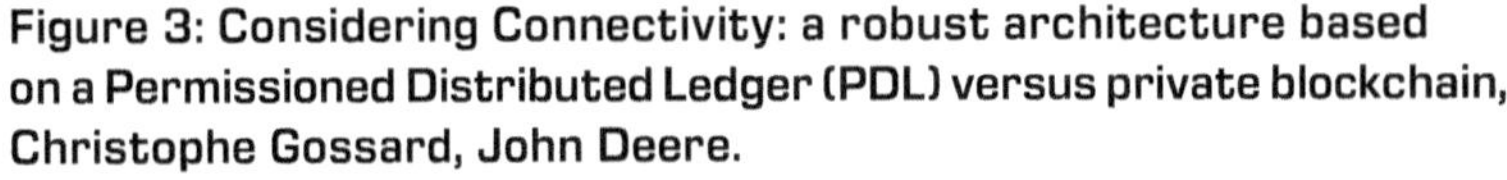

Figure 3: Considering Connectivity: a robust architecture based on a Permissioned Distributed Ledger (PDL) versus private blockchain, Christophe Gossard, John Deere.

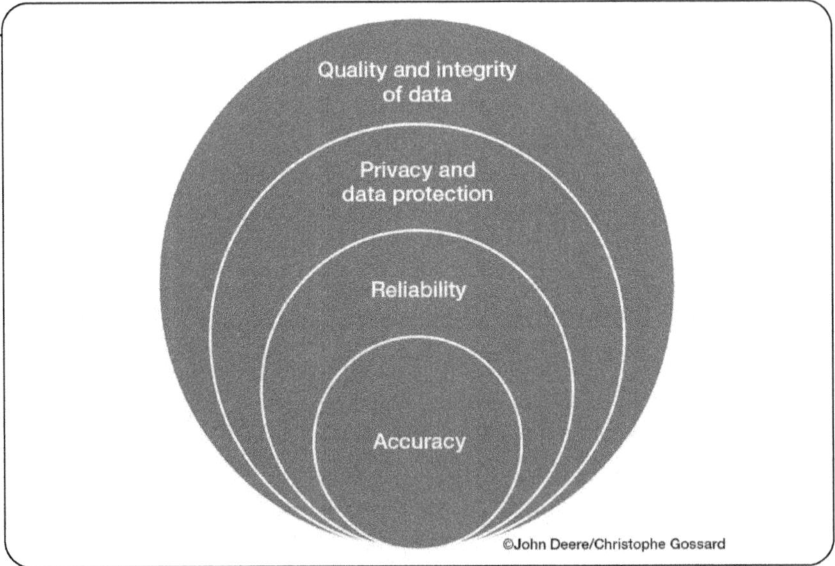

"My company used to say we need to have the feet on the ground and the eye on the horizon," he says. "The tractor of today requires a lot of time, a lot of hours invested, and very little value is provided to the farmer. What we need to survive is for the farmers to have solutions for the future."

▌ It's Possible if Farmers Can Rely on the Machines

At a family-owned farm, nursery and garden center in Le Thillay, France, 18-year-old Antoine Chatelain has taken the lead in bringing robotics into the mix. Chatelain Nursery belongs to Antoine's father, Laurent, but the younger Chatelain manages the technology. He sees driverless tractors as another means of improving farm work.

"There are tractors already using GPS out in the field, but what we are seeing is that not all tasks can be automated with current technologies," Chatelain says. "I think that artificial intelligence and other systems can greatly improve what we can do, and I think, in the end, we will only be required to do only part of the job. The other part of the job will be reduced greatly by AI, autonomous tractors and other vehicles."

Specifically, Chatelain envisions a future where advancing technologies help enhance decision making. The farm will never fully drive itself, he says. Instead, the value proposition of AI will continue to be performing intensive, repetitive tasks, creating space for new robotics-related jobs to develop.

If the current generation of agricultural robotics is any indication, the technology needs to be finessed before a driverless tractor becomes a viable option. From Chatelain's experience, the biggest issue is reliability.

"Sometimes the robot works great, and we don't see any issues with the work," he says. "The robot does its job, and it is a great success. But at a moment's notice, the machine can hurt your crops. You may have some issues that can be attributed, in part, to wrong settings being entered or environmental variables that are not in kind with what the robot can do."

Chatelain says the cost of ownership can also be preventative. Once purchased, however, the systems require little human supervision and are very inexpensive to run.

"For us, driverless tractors and robots are really about reliability and the trust we can put in those systems," he says.

▌ It's Possible if Robot Manufacturers Can Deliver Proper Training

Naïo Technologies CEO Aymeric Barthes believes agricultural robots are already making a massive impact on the farming industry. There is already an abundance of robots that work on vegetables and vines. It's only a matter of time, he says, before driverless tractors disrupt the status quo even more.

"I think the most important thing we have to keep in mind is safety," he says, adding that this is where proper training comes into play.

"When using robots, you need to really train a lot with the robots to understand how they work. to understand how to integrate the robot in the fields, how to set up the robot in the right configuration," Barthes says. "We have technical support integration during the time when you use the robot, so we have a lot of things to do too, and the training is really key."

This means not only training the end user, but also the distributors. When customers understand the technologies, they are more likely to use them and spread the word. Distributors that know how to work the robots operate with a more powerful means of selling the product and alleviating potential fears.

On the issue of safety, driverless tractors have a seemingly quicker pathway to the market than, say, fully autonomous cars. Driverless cars operate in the public sphere. The biggest benefit farms provide on the road to implementation and adoption is privacy.

"Agricultural robots work in fields away from public areas," Barthes says. "When it's possible, the farmer can help create a safe environment by putting barriers and signage around the areas where autonomous machines will be operating. This is very different from driverless cars because you cannot privatize the public area."

In addition to managing potential safety concerns, Naïo Technologies, the innovator behind the Oz, Ted and Dino weeding robots, focuses on profitability. Robots, Barthes says, must be profitable in order to succeed. One way that agricultural machines can increase their value is by addressing the power issues. Driverless tractors will need to address battery-related problems, too, and that's precisely why companies like Naïo Technologies are already thinking ahead.

▌It's Possible if the Technology is Robust

"When you have batteries on the robot, you need to charge the batteries, and if you have to, you have to send someone to charge the batteries, you have a problem," Barthes says. "You have an issue with the profitability of the model. We need to make that aspect of the agricultural robot autonomous like the vacuum robot in the house. So, we designed a trailer with solar panels to charge the robots autonomously."

As the General Manager Solutions for VARTA AG, Gordon Clements cites batteries and power as examples of the important role technology plays in developing effective autonomous machines. After all, the batteries that power autonomous tractors and other agricultural robots need to be durable, intelligent and properly optimized.

"From a power perspective, farming without a tractor driver is already possible," Clements says. "The challenge is to consider the battery and also the charger and the availability of charge.

"The notion of charging a battery is probably the key to success because it's not only in the robot that we need artificial intelligence and machine learning," he continues. "The big challenge is really combining the charger and the battery together and bringing artificial intelligence and machine learning directly to the battery, so that we can optimize the battery and the charger as we go forward."

With batteries, safety is a key concern from the beginning. All of the materials that go into making a battery have safety standards they must pass. Then, Clements says, the components must be put together in a safe way. An automated production line is used to minimize deviations in the product quality and ensure the battery can operate within its safe window.

"The battery, especially in the application of agricultural robotics, is subjected to shock, vibration, pretty harsh environmental conditions, in terms of the weather, rain, direct sunshine," Clements says. "All of these things need to be considered in terms of the packaging and how the battery is actually housed, and provided you get all of these things correct and your vendors able to deliver these things to you, then the safety issues can be managed in such a way that it represents a very small risk to the end user, if any risk at all."

When a battery needs to be replaced, the goal is to replace the modules instead of the batteries themselves. This, Clements says, will enable the farmers to safely change the power sources on the farm without expert assistance. An even better option is to use wireless changing solution. VARTA recently created one with power supply company IN2Power.

"It's much more relaxed," Clements says. "You can manage the charging regime, and if we can learn what the power consumption looks like and what the profile of the power consumption looks like and what the charging regime looks like, we can basically collect that data, and we can optimize them for each individual application or even for each individual farmer."

It's also essential for the batteries and the robots to speak to one another. Not all agricultural robots nor driverless tractors will talk to a standard battery module, Clements says, so a universal gateway is necessary. VARTA has already developed one.

"We can basically implement any protocol that the end customer needs," he says, "and give him a completely integrated suite without having to deviate from the standard battery module and lose the advantages that he gets from having a standard module."

▌ It's Possible if the Driverless Tractors Demonstrate Success

While it can be easy to get caught up in the futuristic fantasy of driverless tractors, Syngenta's Chief Information and Data Officer Greg Meyers thinks the farming industry is already there.

"In many ways, agriculture is quite far ahead of consumers in this space," he says. "I mean, tractors have been steering themselves for 20 years, which is farther along than consumer vehicles are, and for the record, tractors with autosteer steer better than human beings do and with less effort."

Fully autonomous tractors are on their way, but that doesn't mean farmers will immediately embrace the technology. Many farmers, Meyers points out, use their parents' and grandparents' methods.

"Agriculture is a complicated endeavor from the moment you put the seed in the ground, and there's thousands of different things that could stop a seed's ability to reach its full genetic potential," Meyers says. "All those things require several hundreds, if not thousands of decisions that human beings have to make. In many cases, farming is still an art. It's something that relies a lot on intuition, judgment, and experience."

Before the industry has access to a driverless tractor, Meyers continues, there will be many steps along the way. The progression is slow, but the autonomous features will continue to make the current equipment increasingly intelligent. These advancing technologies still have a lot to offer.

"Human beings are not good at managing a lot of variability, they're good at averages," Meyers says. "Farming today is still a lot of averages. You apply still pesticides, and you broadcast a similar thing across the whole field.

"But, in the long run, you'll be able to perceive relatively small differences in some plots of land and potentially be able to do something different plant by plant," he continues. "That's not something a human being has the cognitive ability to do, but it is something that I think computers and data science and, ultimately, machinery, as they work together, can unlock. There will be a completely new opportunity to radically change the way agriculture is done."

Regenerative farming, which is based on maximizing biodiversity in the soil and on a field through various practices, is one example. These practices are time-consuming. Someone who needs to spend a lot of time in a tractor each day won't be able to commit to large-scale changes, even the kind of changes that often appeal to their customers.

"It's important to point out that farmers are very different than consumers because they only have 40 times in their lifetime to get it right and to make a living and to pass along the farm, so they're not going to take a lot of risk on things that are unproven," Meyers says. "I think until there's trust around how these algorithms work, and they can demonstrate success on that particular field, not just in demos and tests, but actually on their particular field, it will take a while for them to be willing to take the production risks associated with relying on computers and algorithms to make decisions for them."

Regardless of their analytical capabilities, however, smart technologies remain imperfect. Mother Nature tends to throw a lot of variables at agricultural robots. Hardware that can effectively brave the elements and overall environment can sometimes be hard to find.

Meyers says these things take time, but some of the very repeatable tasks, like seeding and harvesting, benefit from mature mechanics and AI. Driverless tractors will follow suit. Even the safety factor is coming along.

"Just like in the 1920s, everybody was worried about cars and all the issues about having 800-pound machines moving through the streets and hurting people," Meyers says, "but we navigated that, and we will navigate this as well."

4. Safe Positioning and Image Analysis

As agricultural robots and advancing technologies become increasingly sophisticated, farmers are more apprehensive about safety, reliability and accuracy. The key to addressing these concerns lies in continually working to improve the technology.

In the "Safe Positioning and Image Analysis: Are Technologies Reliable Enough for Autonomous Works in Fields?" industry experts shared how agencies, businesses and universities are working to improve two essential technological capabilities. There are many moving parts, and they all need to be able to work together. This section focuses on what must happen in order for these technologies to be reliable enough to meet the users' needs.

With respect to safe positioning, Tanner Whitmire (Hexagon I NovAtel) shared how positioning accuracy and integrity impact autonomous solutions, while Joaquin Reyes González (European GNSS Agency) explained why global navigation satellite systems (GNSS) are the preferred positioning option for autonomous robots.

On the topic of image analysis, Hajar Moussanif (Cadi Ayyad University) discussed how deep learning modules can improve how robots "see" and analyze various objects, such as crops and weeds. Markus Höferlin (Farming Revolution) continued the discussion by explaining how deep neural networks can be used to train robots to more accurately identify objects and conditions, and thus, improve their overall performances. Barney Debnam (Microsoft) brought the focus back to the overarching question by offering a framework for understanding whether today's technologies are reliable enough for full autonomy in agriculture.

■ The Importance of Accurate Safe Positioning

In agriculture, positioning systems have been around for quite some time. Today's systems are mostly used for small-scale tasks. The advent of agricultural robots and driverless tractors, however, will require technology providers to think a little bigger.

"Positioning systems are built for the primary purpose of creating applied maps, such as seeding, planting, and yield maps, and for guiding a machine left or right," says Tanner Whitmire, sales and business development manager in agriculture at Hexagon I NovAtel. "Over the last several years, the focus has really been geared towards the accuracy of the positioning system. We initially started out with accuracy within several meters and have been able to take it down to decimeter-level accuracy or centimeter-level accuracy."

As the industry transitions to fully autonomous solutions, Whitmire continues, the simple algorithms and hardware used now will need to be upgraded. Position integrity and safe positioning are becoming increasingly important. This requires technology suppliers to decrease the number of accepted errors. Currently, there are three positioning errors related to agricultural applications: positioning error, cross-track error, and along-track error.

Positioning error, Whitmire explains, is one that occurs in longitude and latitude. The next error is cross-track error, where the machine is off to the left or right when attempting to drive down a path. A small cross-track error, Whitmire says, will represent that the machine is close to the midline, but the greater the error, the harder the farmer's job becomes.

The third is the along-track error where the vehicle drives down the path but is too far ahead or behind where it should be. All three of these errors are in relation to horizontal accuracy, but Whitmire believes the future will require technology companies to consider vertical accuracy and other variables, too.

"Position agriculture today is really been heavy focus on the accuracy of the position," he says, "but as we begin transitioning to the fully autonomous solutions, we will need to expand our focus to include the position integrity."

The difference between the two can be a little confusing, but Whitmire describes it in terms of confidence level. In the task position (so-called accuracy), "the accuracy is represented at a 95% confidence level." Meanwhile, in the assured or safe position (integrity), there is confidence at the 99.999th percentile.

"The task position represents a more precise accurate position, as it is accounting for fewer errors," Whitmire says. "It is a simpler software solution that allows us to maintain our level of accuracy. As we expand the integrity of that position, we have to account for more errors, which is going to stimulate the challenge of having to advance our software."

Figure 4: Achieving safe positioning in autonomy, Tanner Whitmire, Hexagon | NovAtel.

In order to move toward a safer, smarter and more advanced system, companies that focus on positioning need to include sensors, GNSS, and other technologies to help improve accuracy.

"And by adding these types of technology and these features, it's going to allow us to increase our confidence to 99.999 percent," Whitmire says. "This is essential as we transition to the fully autonomous solution and take the driver out of the cab. We need to make sure that we are 99.999 percent confidence that that vehicle will not run into anyone or anything."

▌The Importance of Signal Reliability

As the robotics manufacturers and technology companies focus on improving the positioning systems from a hardware and software perspective, many others are focused on the satellite navigation services that help make accurate positioning possible. One such organization is the European GNSS Agency (GSA).

"We are the ones operating and looking for applications for these satellites," says Joaquin Reyes González, marketing development technology officer for GSA. "What we are putting online is the Galileo signal."

Galileo is Europe's GNSS, but is only one of the solutions GSA has developed to ensure safe positioning. The agency also has the European Geostationary Navigation Overlay Service (EGNOS), which helps to improve the accuracy of GNSS, and Copernicus, the European Union's Earth observation program, which is focused on "information services" that draw from satellite Earth Observation and in-situ (non-space) data.

"What is clear is that this satellite technology and all this information that we are receiving is giving us a new generation of farming," González says. "The farmer is not alone."

With Galileo—a system developed 20 years after traditional GPS—users benefit from a better experience. The first thing that sets this system apart is that it's multi-frequency. Users have access to open-service multi-frequency, where messages are delivered using several channels to improve accuracy.

"Galileo satellites are transmitting the message, not only using one channel, but using several channels," González says. "This results in better performance and a better position for the user because the receiver can receive exactly the same information coming from different channels."

Galileo also offers open-service navigation message authentication (ensuring the satellite is coming to the right place), signal authentication service (encryption ensuring the signal is coming from the right place), and high accuracy service (improving accuracy and precision). Combined, these features produce a highly accurate and precise system that users can trust to support safe positioning.

"If I have one message today, it is for users to consider using GNSS and not only GPS," González says. "This is the most important thing."

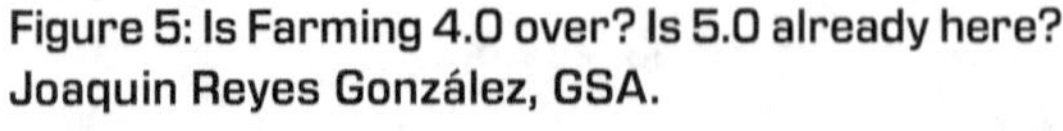

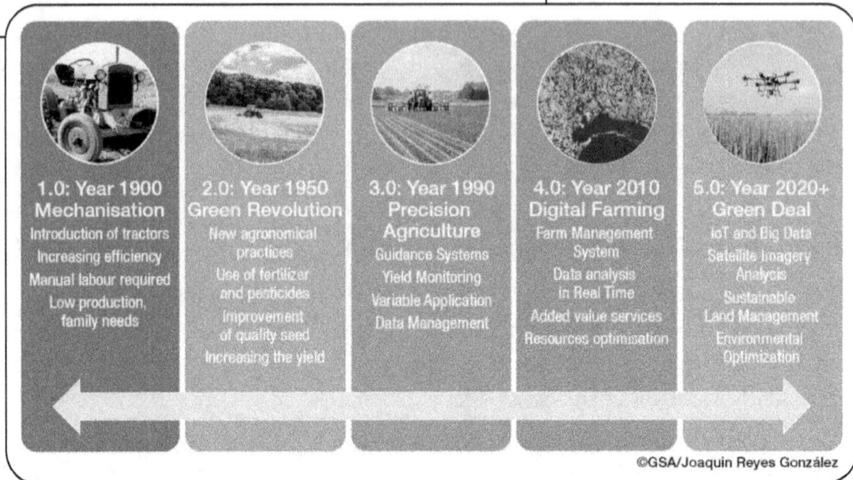

█ The Importance of Deep Learning

In addition to safe positioning systems that help agricultural robots navigate in complex environments, image analysis is equally important for enabling these powerful machines to perform their jobs effectively. Eliminating or streamlining tasks is certainly an important part of democratizing farm work. Hajar Mousannif, an accomplished Cadi Ayyad University associate professor, who founded the Master program of Data Science and is the Golden Winner of WomenTech Global AI Inclusion Award 2020, is thinking even bigger.

"I gave a TEDx talk where I said I was haunted by a dream, a big dream to change this world in some way or the other," she says. "I've always been confident that technology and artificial intelligence have the power to make people happier. In the five years since my talk, the world has indeed changed, but people haven't gotten any happier."

COVID-19 has caused devastating impacts around the world. In Morocco, Mousannif's home country, unemployment, social disparities, and poverty are on the rise. This impacted her outlook.

"It has taught me that making people happier can be achieved through creating cool tech, but it's more about guaranteeing a decent living and ensuring people's basic needs."

Agricultural robots can make a huge difference. According to Mousannif, who is leading a project with U.S.-based company FotaHub, Inc. to expand the capabilities of Shama, the first Moroccan-made humanoid robot, they already have.

"With the massive adoption of technology in the agricultural sector, we're not only allowing agriculture businesses to be more profitable, more sustainable, safer, and more environmentally friendly, but it will also ensure that farmers in rural areas make a decent living and get an education, instead of spending several hours per day doing highly repetitive and physically taxing labor that robots can efficiently do," she says.

"In Morocco, we have succeeded in automating many tasks related to agriculture labor. Thanks to computer vision, many machines are now capable of identifying and sorting different fruits and vegetables according to their size and degree of maturity."

This capability is the result of deep learning, an artificial intelligence function that helps computers and robots mimic the human brain. Specifically, deep learning imitates human intelligence, enabling machines to process information, reason, adapt to the environment, and solve more complex problems. It can be used in agriculture to help robots process images of plants and help detect diseased crops. To be successful, these deep learning modules need to be trained with data.

"We need to feed the algorithm with images that have the type and location of defects or diseases and let the model learn how to locate them after it has been trained," Mousannif says. "It's worth mentioning that defects, whether it is a disease or a pest, can be challenging to detect even by human inspectors. It is also very time-consuming, so anything that is improving automation, increasing efficiency, and maintaining high quality in production is more than welcome."

When attempting to train the deep learning modules, environmental factors will create challenges. The algorithms must be able to account for things like sunny and cloudy conditions or plant shadows. Images with insufficient light or improper exposure impact accuracy. Depending on the application, other image processing techniques can be used. An image can use sharpness to remove blur from some images, for example. Depending on the context, things like blur can be detrimental to a machine's accuracy.

There are some technologies that can help. Convolutional Neural Networks (CNN) are trained to classify images in the same way than humans do. Transfer learning is a machine learning method that solves the problem of not having enough or any labeled images in the data set. Image segmentation can be used to classify, detect and segment anomalies. In the agricultural world, this means determining weeds and diseased plants among healthy plants. There are many people contributing to this important work, but the journey to super accurate autonomous technologies is a long one. Mousannif remains focused on the end goal.

"By actively contributing to the AI field, we will not only ensure a better future for us, but also for other generations to come," she says. "We need to put our efforts toward building tech that really matters, that solves real problems, that has a direct impact on people's lives and, most importantly, that ensures inclusion and preserves our humanity."

▌ The Importance of Improving Computer Vision

Many of the most important farming tasks are relatively undesirable for humans, which is why the mass adoption of agricultural robots, as Mousannif says, will impact people's lives in a big way. With its AI-powered robots, the "weeding as a service" supplier Farming Revolution (formerly Deepfield Robotics) is focused on a particularly tedious and often-dangerous job: weeding.

The robots drive autonomously through the field, using multi-spectral cameras to take images of what they encounter. Then, using artificial intelligence and deep neural networks, the robot identifies the crops and weeds. The weeds closest to the crops are then effectively removed without chemicals. It sounds simple, but Markus Höferlin, head of artificial intelligence at Farming Revolution, says the end result is the product of an incredible amount of behind-the-scenes work.

"In the beginning, one of the main challenges was the computer vision part," he says. "So, why is this so challenging? If we look at some industry plants where we also have a lot of computer vision tasks, we always see that we must constrain our environment to be able to reduce the complexity of it."

A farm field, however, is completely unconstrained. There are different soil types and crops with leaves, and some of the crops overlap with weeds and with one another. The plants also tend to look different in the morning than they do at night and from one day to the next. An ever-changing outdoor environment further complicates the problem. Farming Revolution went to work.

"We are actually putting immense effort into capturing and labeling data in all the varieties that we can think of," Höferlin says. "We captured data from more than 50 different fields. We started our capture campaigns at two o'clock in the night, where it's completely dark and the plants are sometimes still closed, and in the morning when the plants open slowly. We capture images into the afternoon and until the evening begins. So, we have all the different light conditions that we can think of."

Farming Revolution also captured data in various weather conditions (dew, dust, mud) and seasons. The data was collected for five years. More than 65 different species were labeled with 99 percent per-pixel accuracy. The result was more than 12 million annotated images—a huge amount of training data. The artificial neural network would use this data to help the robot analyze and process the field images.

"This is what really makes a difference," Höferlin says. "With this variety of data, we managed to go through the field, turn the robot on, and go for it. We were able to do this on an accuracy level of 99 percent, without needing to retrain the neural network and overcome the problem of generalized data."

Even with this amount data, however, Farming Revolution is continually working to capture more. The more data, the better the insights gained from qualitative analysis. It also enables the company to troubleshoot any problems and improve the network over time.

"If a farmer has our robot, for example, and sees that on one particular field, he is not happy with the results. He can tell us, "please check what happened or make it work better here"," Höferlin says. "The farmer can just upload the data to us. We can conduct evaluations on this data and see where we might have some problems, and if we see that, for example the classifier underperforms, we can label the data according to what we learn and use this data in our training set as well."

█ The Importance of Continual Improvement

Safe positioning and image analysis are improving every day, but the biggest question remains: Are the technologies accurate enough to reliably work autonomously? Barney Debnam, director of agribusiness solutions for Microsoft says it depends.

"It depends on the problem that you're attacking with autonomy," he says. "It depends on the risk profile of that particular agricultural operation."

Debnam describes the three main risks profiles that need to be accounted for: risk to humans, risk to crops or livestock, and risk to environment.

"Thinking about these risk profiles is particularly important for the companies developing solutions", Debnam says. A robotic system that follows a picker and provides a basket to carry the commodities, for example, is a different scenario than a sprayer that may apply a pesticide in a vineyard above a water area that delivers resources to the public.

"For all of us who are designing the autonomous systems and thinking about risk, it's important to qualify the risk and understand the impact," Debnam says,

There are several frameworks for assessing whether the technology is robust enough to deliver confident, safe outcomes. Debnam specifically names the American Society of Agricultural and Biological Engineers, The Journal of Agricultural Safety and Health, the National Institute of Standards and Technology, and the ENISA tool from the European Cyber Security Act. He encourages technology providers to use these resources. This is one of the ways technology suppliers can build trust with their customers.

Another way is to address what Debnam calls "basic surveillance." This is a desire from food companies and consumers to get more transparency. A lot of the solutions are focused on performing a task, rather than proving information that a certain production process happened at a specific time on a specific day. When robots become fully autonomous, customers will expect more clarity on how they work and what they do.

The goal is to create a situation where everyone thrives. Mousannif expresses this best:

"We can use artificial intelligence, and we can use technology, but what's the use in tech-nology if we cannot create value, and we cannot make this world a better place to live?" she says. "AI and technology in general have to go hand-in-hand with social impact and value creation."

5. From Lab to Success Story: What Business Model?

There has never been a more exciting time to be a part of the agriculture industry. The second industrial revolution continues to deliver advancing technologies that previously existed only as pipe dreams. As plans become prototypes, and successful prototypes become trusted tools, agritech companies and investment firms look to the future. Perhaps the biggest question remains: What happens next?

In the "From Lab to Success Story: What Business Model?" keynote, IDTechEx technology analyst Michael Dent shares market research that hints at the path ahead. After his presentation, six well-informed panelists continued the discussion.

Seana Day (Better Food Ventures & The Mixing Bowl), Peter Hanappe (Sony CSL), David Bowles (The Yield Lab), Erik Pekkeriet (Wageningen University & Research), Sarah Waltner (Raven Applied Technology), and Daria Batukhtina (Kubota Holding B.V. - Innovation Center) debated which agricultural robot development and business models are the most successful.

This section, not only details where the industry is headed, but also what's happening right now. One way to assess the current state of the agrorobotics sector is to understand which technologies are being funded right now. In the "Start-ups & Investors - How to Partner Effectively" session, two start-up CEOs (Naïo Technologies' Aymeric Barthes and EcoRobotix's Aurélien Demaurex) and two investors (Capagro's Tom Espiard and BASF Ventures' Dr. Claus Hackmann) share their experiences. These four panelists explain how each party can best collaborate to move everyone into the future.

▌ The Key Technology Trends in Agricultural Robotics

Michael Dent began his keynote by stating familiar facts: The population is growing but food production is not. In addition to the need to increase food production by 70 percent, another potentially troubling trend is that the number of people working in agriculture is decreasing. Labor costs are increasing across the industry, which is squeezing margins on farms.

"Despite the clear need for digitization, agriculture has historically been fairly slow to adapt, often lagging behind other Industries in terms of adoption of digital technologies and robots," Dent says. "However, over the past five/six years or so, this has really begun to change. Farms are adopting more and more digital technologies, and increasing numbers of startups get funded across the world, as they try to take advantage of a skill gap in the market."

A recent IDTechEx report, "Agricultural Robots, Drones, and AI: 2020-2040: Technologies, Markets, and Players," identifies six areas that will have the biggest impact on the

Figure 6: Robots and drones: market and technology readiness by agricultural activity, Michael Dent, IDTechEx.

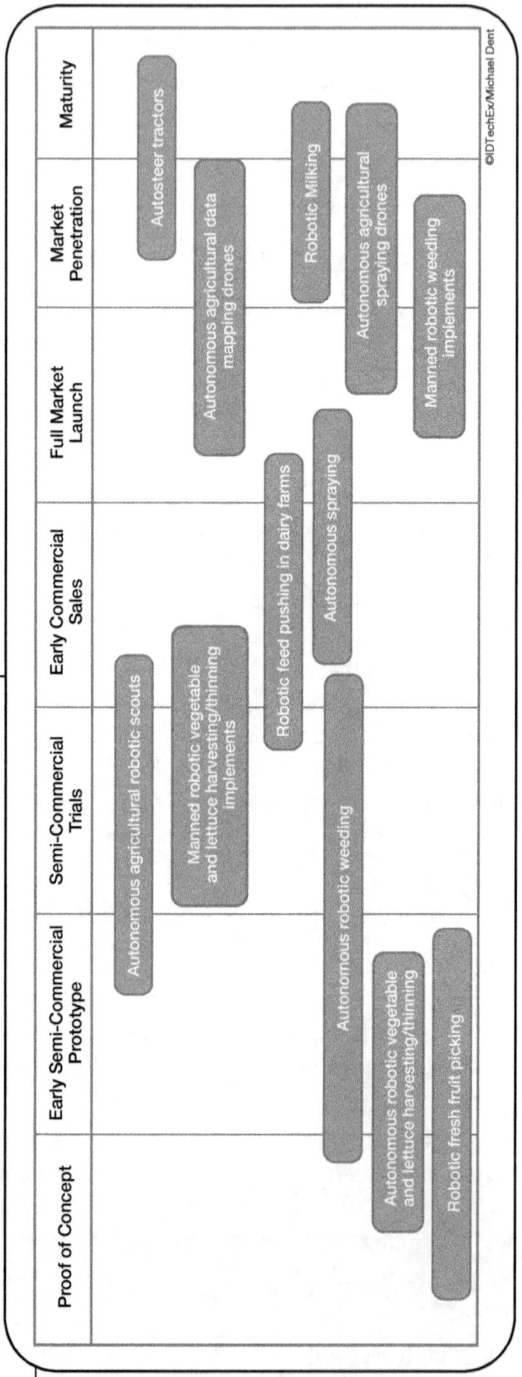

industry overall: small autonomous robots, intelligent tractor-pull implements, robotic implements with simple vision/control, autonomous tractors, autonomous sprayers and fresh fruit picking.

"At IDTechEx," Dent says, "we believe that the uptake of precision enabled agriculture could significantly disrupt the existing agricultural supply chain."

In the future, he continues, farming will become more reliant on and driven by data. This will create additional value along the supply chain, from sensor and robotics equipment manufacturers to data analytics firms. The face of agriculture is certainly changing, and IDTechEx has identified a few key trends that are helping to drive this transition. One such trend is toward ultra-precision agriculture.

"Constant-rate technology, where entire fields are managed as one entity and where inputs, like nutrients, water, and crop protection products are applied uniformly across the field with no site-specific customization," Dent says, has started to transform to variable-rate technology.

This helps to significantly reduce the inputs used and increase yields because parcels of land are managed based on their unique needs and the treatment required. Technologies like geolocation, sensors, drones collecting data and images, and more are used to assist with variable-rate technology.

"This trend will continue over the next few decades, evolving from variable-rate technology to ultra-precision agriculture," Dent says. "And in ultra-precision agriculture, the whole farm is managed on an individual plant basis, where the exact doses of inputs are applied to each plant based on its measured needs."

The industry is in the early developmental stages of ultra-precision agriculture, but IDTechEx estimates that it could arrive within the next two decades. Another key trend will develop in a similar way. Dent cites the use of artificial intelligence for better disease prevention, yield prediction and quality management.

This is rarely used now, Dent says, but this will change. The industry will begin with simple AI and move toward more complex AI systems. Ultimately, as the algorithms are improved, neural networks will become more common, and IDTechEx estimates that animatronic robots will enter the industry within the next 20 years. Dent says these technology trends will be supported by three enabling technologies: vision technology, AI and autonomy.

These enabling technologies will eventually support robots and autonomous machines, but they too are in the earlier stages of development. Computer vision, for example, came a long way in the last decade, Dent says, but the computationally heavy algorithms still require a lot of energy to run. In time, AI will go beyond computer vision to plant recognition and yield prediction, all of which helps to accelerate the transition to ultra-precision agriculture.

With autonomy, Dent says, the challenge is finding the balance between the cost of additional sensors and their benefits, as well as combining a variety of technology systems into one machine that can, for example, see, spray, and navigate the challenging landscape.

"The increasing drive toward autonomy is likely to cause a long-term paradigm shift in agriculture towards smaller agricultural vehicles, with robots becoming increasingly cost competitive," Dent says. "We think that the transition towards small autonomous fleets of robots, as opposed to single large pieces of agriculture machinery that are operated by a single user, will lead to a fairly significant shift in the way farms are run over the next two decades."

▊ What Makes a Successful Robotics Development and Business Model?

While IDTechEx is busy researching the upcoming trends, others are working to bring them to life. Things continue to move quickly. When it comes to outlining the right ingredients for a productive and profitable agtech venture, however, it's clear that everyone uses a different recipe.

For Sony CSL researcher Peter Hanappe, the recipe includes a lot of teamwork. Hanappe works primarily with small, mostly organic urban farms, creating advanced robotics tools for agroecology, the study of the relationship between agriculture crops and their environment. One of these tools is a small, lightweight weeding robot. The work of creation is collaborative by nature, but Hanappe works in such a way that invites others to join the process.

"We always said we would make the hardware open source because we're from computer science backgrounds, and we have a tendency of working open source," he says, adding that the team wanted to explore alternative business model, and open source seemed like an intriguing option.

"It's probably the idealistic maybe a bit naïve," he says. "We still have to show that it's possible. So we're not sending anything yet. The proof is in the pudding, can we really build a business model on top of that? I think there are good signs.

"In one sense," he continues, "I think farmers like to know the technology that they're using. I mean you could completely offer these products as a closed box or even as service maps, but I think it's nice also to explain to the users how it works. Another thing is if you want to be able to collaborate with, say, scientists and engineers and farmers and create a community around that, then I think going open is the way to conquer peoples' hearts."

Weighing in on the investment side, Better Food Ventures & The Mixing Bowl partner Seana Day believes describing the right business model is complex. It begins with understanding adoption rates based on crop/commodity, and the size and scale of the operations within the sector where the robotics company is attempting to create a solution. Add to that the tech company's internal development, milestones, inflection points, and where they are in the proof-of-concept/commercialization/scaling up process.

"It's complicated, and then of course we have to overlay that with some of the external factors," Day says. "Is there adequate support for those technologies in the field? Is there a field service organization? Are there boots on the ground? Is there sort of an external service provider that's able to maintain and help support those technologies?"

According to Day, investors also evaluate companies based on the potential route to the market. Having the capability and support to get there is essential. Right now, most investors are focused on funding smaller projects.

"We're still decades away from a billion-dollar start-up robotics company," she says, adding that things are progressing. "This has been a breakout year for farm management software and some of the more software-oriented business models have gotten traction and have gotten past the commercialization. Now, they're starting to see some market penetration. It's just going to take time."

David Bowles, a venture capital investor from The Yield Lab, is also seeing a lot of activity in the agtech space, but he believes a successful business model requires differentiation.

"What's very frustrating is a lot of these companies are solving the same problem, and what I mean by that is a lot of them are solving the problem of navigation around fields or having a communications platform," he says. "To some extent, they take very basic problems, and on top of that, each one adds something new and unique—a new way to detect diseases in apples and a new way to weed a field of carrots.

"So, it's something interesting and unique for each these companies," he continues, "but they're also having to do a huge amount of work just to get unique technology on the platform. For us, it's really frustrating because of the huge amount of time resources wasted, and we're kind of interested in more platform technologies and people that can bring their unique insights, unique capabilities and leverage existing technologies, automation, or navigation devices."

For this season, the agtech space is mostly filled with medium-sized companies that have been funded. Larger companies are few and far between. Companies that get funded are the ones who can produce a venture capital return, and those, Bowles says, are the ones that offer a large number of different technologies and feature sets. He warns against becoming too invested in the nitty-gritty technologies by referencing two companies that have been dominant within the agriculture industry for many years.

"John Deere did not become huge purely from weeding," he says. "Case IH did not become huge purely by being a seeding company."

One way to develop a wide range of technologies is to work with others. Based on his experience as the program manager agro food robotics at Wageningen University & Research, Erik Pekkeriet understands the way participating in research and development (R&D) can support a viable business model.

"Robotics demands a different kind of R&D that is must more intensive, but the payoff can be very good," he says, adding that many companies are hesitant to invest in R&D, particularly when there isn't room in the budget, and they don't see the value. Sometimes the best way to do research is through active collaboration.

"I think the Dutch are, at least, very open to community, and they communicate everything with each other," Pekkeriet says. "So, it's a low step. It's very familiar. And I think that's also a good point of agriculture and food. When you talk to farmers, it's easy entrance. You can start testing immediately. And I think that it's good to have an end user."

Daria Batukhtina, business development manager for Kubota Holding B.V. - Innovation Center, couldn't agree more. She believes focusing on the farmer is the only way to ensure successful solutions and mass adoption.

"Kubota wants to go beyond our business domains to provide total solutions that support the food value chain, and this type of the digitalization path is not possible without knowing the real near- and long-term needs of the growers and farmers," Batukhtina says. "It's not possible without cooperation."

A successful business model also goes beyond what end users need. It also requires companies to support them and play a role in their future profitability. After all, there's a lot of increased pressure on farmers to produce more and produce sustainably, all while reducing carbon emissions and saving the planet.

"If we consider the historical aspect of this, for about 400 years, the farmer was the most important person on the planet. How do they feel today?" Batukhtina says. "This is a collaborative work. This is a joint effort. You can't develop the sustainable business model if you do not consider the life circle of the farmer."

Ultimately, the best business model focuses on collaboration. Sarah Waltner, general manager at Raven Applied Technologies, says it best:

"No one company. No one University. No one agency is going to bring everything that is needed," she says. "I think it's important to really ask 'what do we need to have direct control over?' Is it direct development or composition? What things are not our strong suit or aren't a part of what we bring to market? Then, for those things, we rely on collaborative partners."

When everyone works together, every step of the process becomes a bit easier. Each partner can operate in their own zone of genius, the place they are best equipped to provide expertise.

"We are really focused on helping the actual farmer with the fundamental needs to be profitable, to be sustainable, to know that they can get their job done on time," Waltner says. "That really is our striking force."

▌ Startups and Investors: How to Partner Effectively

It's important to note that there's more than one successful business model. Collaboration is key between manufacturers and farmers, but it is also essential in the relationship that exists between robotics start-ups and their capital investors.

The startup-investor relationship is often described like a marriage. There are those that blossom into a beautiful partnership, and those that were doomed from the start.

To get the inside scoop on what works, Aymeric Barthes (Naïo Technologies) and Aurélien Demaurex (EcoRobotix) shared their experiences as startups working with investors to fund their robotic solutions. Alternatively, Tom Espiard (Capagro) and Dr. Claus Hackmann (BASF Ventures) gave their perspectives as investors supporting the next big innovations on the path to profitability.

Based on their discussion, effective partnership requires five key components. Here's what they had to say.

Trust

"One of the things that is really key in the relationship for me is trust," says Demaurex. "It's the number one. When you work with investors, of course, there is a honeymoon at the beginning of from both sides. So, you have to trust each other. The investor must trust the startup because they are investing in it. There are a lot of things that can lead to mistrust that's created on the startup side. For example, not delivering according to the plan. Changing strategies is quite common in startups."

Communication

"It is important to talk frankly and face reality," Barthes says. "The reality of the startup is that it's violent. Everything is moving very, very fast, and every day, you have to face a new issue, a new problem. You're concerned about everything from your clients to the legislation to the next fundraising operation. So, it's really important to be able to talk.

"I am the CEO of a startup, so the team I have in front of me is the team of investors, and I'm expecting them to really be able to face the reality and to understand our problems and to bring solutions based on their experiences to help us to make our vision possible," he continues. "That is the main problem we have, so it's important to be reactive. Sometimes, investors are too much based on their experience and not enough tuned into the future.

"I think the biggest difference between startups and investors is that we believe in the future and the new way to manage the company, the new way to manage our issues. On the side of the investors, they are very experienced, and they make decisions based on that. It's important to have both sides. I don't want that to change, but it is why we need to find the right words."

Shared Values

"In the beginning of the investment, you should have shared goals and shared values with the company, and I think one of the first steps is to have a strategy day," Hackmann says. "This way, the investors and the management of the startup can get really aligned on the strategy, make a deep dive to have a common understanding about the go-to-market strategy and also on the exit. If you are aligned on that, basically everything else will follow."

"As in a good marriage, you still can learn from one another over time," he continues. "Corporate can learn from startup new technologies, new dynamic, new applications, new approaches. From the other side, startups can learn how to basically find a strategy and use a network and how to execute. Give the startup a platform to execute their new technology and showcase it and run it, and you'll work together for a longer time. This is to give them a boost to go into the right direction."

A Good Contract

"You need a good contract. We are in 2020, it might not last," Espiard says. "There needs to be a precise understanding of what is the exit. You want to make sure that you build value together, but at some point, in the life of startup, there will be an exit scenario. This has to be understood by both parties. It needs to be worked out in governance so that as an investor, I don't impair the value for each of us with an exit and challenge the exit scenario and balance."

"Most venture capitalists (VCs) have an investment horizon," he continues. "That investment horizon is 4-5 years generally because the VCs have investors themselves and have to give back the money to the investors. So, when we make an investment, we have to have an exit strategy in order to liquidate and give that money back to our investors. In the governance, there is a clause that will say, at one point, you will enter into a process where the company will be traditionally be sold. There are some ways to balance the right of the corporation in case they want to have a say and make sure the company and the process is not jeopardized."

A Shared Vision

"I think the key is to maintain a recurrent information flow," Barthes says. "We choose to call all our investors very often, show them our strategy and build the strategy with them as much as possible because the more you have investors around the table, the more complicated it is to synchronize the strategy and build the strategy. You have to take the time to do that. It's really the key to ensure a good relationship because if they share the strategy with you, if they share the vision it's easier to take the necessary steps to make it a very successful partnership."

CONCLUSION

The FIRA 2020 event was one to remember. The first virtual meeting in the event's five-year history, FIRA 2020 proved that when people around the world come together to share their knowledge and insights, great ideas prosper.

Attendees learned from farmers who are thriving in the face of challenges and increased pressure. They heard from robotics manufacturers and advancing technology suppliers that are working tirelessly to support farmers' efforts with new innovations. VCs shared their vision for the future, while industry experts—ranging from robotics lawyers and government officials to those managing agriculture cooperatives and organizations—discussed what must happen for mass adoption to further transform the farming sector.

Over the course of three days, four keynotes, nine workshops, 15 robot demonstrations and endless conversations in formal roundtable events and informal online chat rooms, the industry's most pressing issues and most profound successes were on full display.

In Chapter 1, Food & Farming, presenters set the stage for understanding why agricultural robots have been crucial in helping to overcome labor shortages and address the problem of food insecurity. During a year that revealed and reflected these present and future issues, the speakers were able to offer their perspectives about the current situation and recommend multi-faceted solutions. New initiatives and valuable partnerships provide a promising view of the path forward.

Chapter 2, Society, "robot bashing" offered an opportunity for companies, organizations and individuals in the agriculture sector to discover what producers truly desire from their autonomous machines. One thing is clear: The end user needs to be a part of the agriculture robotics development process. Technology companies that want to increase market adoption must listen to farmer needs and deliver, sometimes going beyond the offerings they are prepared to provide today.

From Chapter 3, Technologies, experts explained the intricacies of implementing driverless tractors in an industry that is ready for them. While some reliability problems and technological barriers must be removed, the key factors for introducing autonomous machines into the field—training, collaboration between manufacturers and suppliers, etc.—are already in place. The discussions and debates in this section helped to get everyone thinking more critically about what lies ahead.

In Chapter 4, Technologies Continued, the presenters shared details about how autonomous technologies are evolving and improving. Safe positioning has benefitted from GNSS and the introduction of more advanced sensors. Image analysis has advanced by incorporating big data, artificial intelligence, and machine learning into the current frameworks. The result is a wide range of technologies that are better prepared to help agriculture robots work safely and efficiently.

Lastly, Chapter 5, Go to Market, celebrated the many essential relationships that drive innovation—all of which are equally important to progress. Every agricultural robotics project must be developed in partnership with other key players of the value chain. Trust and collaboration ensure these projects remain viable. From the farmers and technology companies that collaborate on robotics that will change the agriculture industry forever, to the start-ups and investment firms that work together to transform these robots from mere ideas into workable prototypes, no one business model is better than another. Everyone has a role to play.

Overall, FIRA 2020 provided a holistic view of where the industry stands during a year of unanticipated upheaval. Some solutions are closer to fruition. Others remain on hold as the world slowly adjusts to a "new normal." Through it all, farmers continue to work, growing and raising the food we eat every day. Around them, the industry will continue forward, improving and advancing along the way.

Many of the same issues that were examined in this year's event, will be delivered from new vantage points and discussed from new perspectives at FIRA 2021. At the next event, we will invite major agricultural machinists to present their autonomous tractors. We will go deeper into the industry's essential technologies, such as data collection and machine learning. Finally, we will try to understand the importance of developing autonomous machines that are easy to use, and that meet or exceed the expected user experience.

We invite you to join us as we dive deeper into these critical topics and many more next year. Sign up today at www.fira-agtech.com/register.

Robot Manufacturers' Information Sheets

Autonomous Sprayer, by Automato Robotics

About Automato Robotics

Country: Israel

Date of creation: 2018

Number of collaborators: 5

Market target: Selling Autonomous Spraying services by July 2021

Level of development: Testing phase

Core business:

Automato Robotics is helping to solve the international labor crisis in agriculture by providing affordable robots to every farmer.

The cornerstone of our solution is a robot platform that can drive in off-road conditions and operate in passive greenhouses (90% of the market). This platform is a game-changer robotic solution set to revolutionize global greenhouse agriculture by creating affordable autonomous farming for the mass market. It will serve as the enabler for greenhouse robotic applications to support the growth cycle of multiple crops.

Our first application is an Autonomous Sprayer attachment, allowing us to make an impact on multiple crops and keep people away from chemicals.

Main partners: Raz Sprayers, Kaufman

Contact: Dror Erez +972507103331

Website: www.automatorobotics.com

About Autonomous Sprayer

Number of robots in service: 1

Main functionality: The Autonomous Sprayer is capable of autonomously completing a spraying mission in an entire greenhouse, needless of dedicated infrastructure.

Features: Autonomous mapping, driving, navigating, spraying, refilling

Size: 1.3 m x 0.73 m x 2 m without trailer

Weight: 90 kg + trailer 100 kg

Productivity: Spray a 1 hectare greenhouse within 3 hours

Price: €30,000

Embedded technologies: Visual slam, GPS RTK, Tensorflow object detection

Energy: 48V DC

Next phase: 2G platform and robot, for mass production

Additional Information: The only Robot for passive greenhouse environments (90% of market).

AutoPicker, by TerraClear Inc.

About TerraClear Inc.

Country: USA
Date of creation: Fall 2020
Number of Collaborators: 24
Target market: Row crop farmers with rocks in their fields, 1,000 acres or more
Level of development: Testing phase
Core business: End-to-end system that focuses specifically on automating rock removal

Main partner: Hillco Technologies
Contacts:
Vivek Nayak, VP Engineering:
vivek.nayak@terraclear.com
Trevor Thompson, President:
trevor.thompson@terraclear.com
Website: www.terraclear.com

About AutoPicker

Number of robots in service: 6 (+12 by April 2021)
Main functionality: The AutoPicker attaches to a utility tractor or skid-steer. Each version has a different degree of automation, from the joystick operated tool used in Fall 2020 up to the fully autonomous solution in development today.
Features: Acquiring rocks. The end-effector mechanism is novel and has been designed to be effective over the widest range of rock sizes and field conditions, while causing minimal soil disturbance.
Size: n/a
Weight: n/a
Productivity: 300-500 rocks/h depending on the rock density
Price: $150/h

Embedded technologies: Computer vision, Deep learning, Robotic manipulation, Electro-hydraulic control
Energy:
The rock requires peak power of 35hp/27kW for actuation which is currently obtained by the auxiliary hydraulic supply. The electro-hydraulic valves require minimal power using a 12V/6A connection from the auxiliary electrical supply.
The primary vehicle diesel engine is used to generate the auxiliary hydraulics and electrical supply.
Next phase:
Compatibility with a wider range of farm vehicles.
Fully autonomous rock picking robot integrating a mobile base with a rock picking system that can be deployed in the field with little or no operator intervention.

AVO, by ecoRobotix

About ecoRobotix

Country: Swiss
Date of creation: 2014
Number of collaborators: 35
Market target: Europe, World
Level of development: Testing phase

Core business: Ultra-precision plant analysis
Main partners: n/a
Contact: www.ecorobotix.com/en/en_contact-2/
Website: www.ecorobotix.com

About AVO

Number of robots in service: 10
Main functionality: Ultra-precision spraying

Avo performs autonomous weeding operations in plane fields and row crops. Using cutting-edge machine learning, the robot detects and selectively sprays the weeds with a micro-dose of herbicide. The centimeter-precise detection and spraying reduces the volume of herbicide used by more than 90%, while ensuring crops are not sprayed for yield preservation.

Features:

- 52 nozzles spot-spraying ramp
- Working Width: 2 m
- Speed: 1 m/s max

Size: 3.75 m x 2.45 m x 1.3 m (L x W x H)
Weight: ~750 kg (fully loaded), with 60 kg batteries and 120 L herbicide mixture
Productivity: Daily Throughput and Autonomy

- 5 ha with 8 hours of autonomy / Up to 10 ha (including night-time operation and battery swap)
- Up to 95% herbicide saved (depending on weed density)

Price: In development / Less than €100,000
Embedded technologies: n/a
Energy: Solar powered with rechargeable batteries. Can treat up to 10 ha/day.
Next phase: Commercialization

Bakus, by VitiBot

About VitiBot

Country: France
Date of creation: 2016
Number of collaborators: Approximately 50
Market target: Winegrowers
Level of development: Marketed
Core business: Design and production of Agricultural or Viticultural machinery

Main partners: VitiBot counts among its partners wine market players such as Champagne Roederer, Martell Mumm Perrier-Jouët or Champagne Laurent-Perrier, Piper-Heisieck and Charles Heidsieck. VitiBot can also count on BPI France.
Contact: Bernard Boxho – Managing Director
Website: www.vitibot.fr

About Bakus

Number of robots in service: 15 to 20
Main functionality: Bakus is able, in time, to carry out most of the work in the vineyard, soil work, spraying, pruning.
Features: n/a
Size: 3.50 m x 1.75 m x 2 m or 3.50 m x 1.95 m x 2.50 m (L x W x H)
Weight: 2,350 kg
Productivity: n/a
Price: Start from €118,000
Embedded technologies: 2 RTK GPS for high precision navigation, 2 inertial control units, 12 mechanical detectors distributed around Bakus, multiple cameras for safety, 4 Lithium Ion batteries for 100% electric power, a range of modular tools that can be easily and quickly installed on the tool holder pole, Michelin Multibib 320/65 R16, 4 brushless electric motors, 4 planetary gearboxes integrated in each wheel, 4 steering motors, electric and independent
Energy: 100% electric
Next phase: Industrialization, export, reliability
Additional Information: By choosing Bakus, you make the choice of sustainable viticulture.

BIPBIP, presented by CTIFL

About CTIFL

Country: France
Date of creation: 1952
Number of collaborators: 280
Market target: Fruit and vegetable sector
Level of development: n/a

Core business: Applied research in the French fruit & vegetable sector
Main partners: IMS, LaBRI, Elatec, Fermes Larrère
Contact: marine.louargant@ctifl.fr
Website: www.ctifl.fr

About BIPBIP

Number of robots in service: 1
Main functionality: Intra-row weeding
Features: BIPBIP is based on a hoeing tool-block that is independent from the carrier and automatically guided by imaging and telemetry, covering a single crop row. It uses a computer vision system to detect row and crop locations and a general control system for mechanical weeding.
Size: n/a
Weight: n/a
Productivity: n/a
Price: n/a
Embedded technologies: The vision system is composed of a RGB camera and led panels to control light conditions, a row detection algorithm to guide the module along the line and a crop and stem detection algorithm (based on neural network).

Using crop and stem location, the control system can activate different tools along and into the rows, hoeing weeds around the crops. The hoeing system is composed of different tools: a mechanical rod and a small ploughshare, to be compatible with different kinds of weeds or crops.

Next phase: Currently, BIPBIP is adapted to maize and beans. It can be adjusted for leek, onion, and beet productions.

Additional Information: BIPBIP combines the imaging (IMS), robotics (LaBRI) and mechanical design (Elatec) skills necessary for the development of the project. It is backed by the professional expertise of a market gardener (Fermes Larrère) and the agronomic and experimental skills of a technical centre (CTIFL) required for producing specifications, performing tests and carrying out agronomic validation.

BIPBIP project is supported by Agence Nationale de la Recherche.

CEOL, by Agreenculture

About Agreenculture

Country: France
Date of creation: 2016
Number of collaborators: 45
Market target: Agriculture
Level of development: Marketed
Core business: Agreenculture is a French company that designs, develops, and produces autonomous solutions. With 13 years of Research and Development in the field of positioning and satellite guidance, we have been investing this know-how in the development of robotic solutions for the past four years.

Partners: We share this passion with our partners and collaborators (Dassault Systèmes, Kuhn, Pellenc, Région Occitanie…) working on innovative and economically viable solutions.
Contact: contact@agreencutlure.fr
Website: www.agreenculture.net

About CEOL

Number of robots in service: 15
Main functionality: CEOL works autonomously, without pauses or human assistance. In addition to its accessible mechanics and maintenance, CEOL is equipped with a range of options that allow the user to adapt the machine to his needs. The easy-to-use machine management interfaces allow you to maximize the robotized surface on your farm.

Features: CEOL is a tracked robot, equipped with a standard implement hitch allowing it to interface with a wide range of agricultural implements. It can carry or tow a range of equipment of different sizes: Shredder, Narrow frame, Large frame, FACA roller and Crumbling disk, kress fingers and inter-vine hoes.

Size: 170 cm
Weight: 750 kg
Productivity: 1 ha/h
Price: €50,000
Embedded technologies: GNSS RTK positioning, (centimetric precision)
Energy: Hybrid
Next phase: Commercialization in 2022

ChickenBoy, by Faromatics

About Faromatics

Country: Spain
Date of creation: June 2016
Number of collaborators: 19
Market target: Broilers
Level of development: Marketed

Core business: Artificial intelligence, big data, robotics
Main partner: Big Dutchman
Contact: sales@faromatics.com
Website: www.faromatics.com

About ChickenBoy

Number of robots in service: n/a
Main functionality: 24/7 automated surveillance of chickens
Features:
- Monitoring excrements for early detection of changes in faeces related to e.g. health problems
- Detection of:
  - Mortality in the path of the robot
  - Wet spots in the litter (in preparation)
  - Defective nipple drinkers
  - Measurement of free usable space (a measure of growth and distribution)
  - Maps of thermal comfort (temperature, relative humidity, air speed), air quality (CO_2 and ammonia), light and sound

- Light-induced movement of birds when necessary (in dependence of free usable space; manual mode is included)
- Streaming video and audio
- Photos of animals and equipment at any interval
- Alarms (email, SMS) for user-defined anomalies in the shed

Size: n/a
Weight: 28 kg
Productivity: Savings of 12-24 m€ per 1,000 sheds
Price: n/a
Embedded technologies: n/a
Energy: Batter powered with autonomous recharge
Next phase: Rollout
Additional Information:
https://youtu.be/ymsiaWFwmYY

Dino, by Naïo Technologies

About Naïo Technologies

Country: France
Date of creation: 2011
Number of collaborators: 70
Market target: Agricultural Robotics
Level of development: Marketed

Core business: Naïo Technologies designs, manufacturers and markets agricultural robotics solutions.
Main partners: n/a
Contact: contact@naio-technologies.com | +33 9 72 45 40 85
Website: www.naio-technologies.com

About Dino

Number of robots in service: 45
Main functionality: Mechanical Weeding for vegetable crops
Features:
- Inter row mechanical weeding
- Robot which navigates autonomously
- Helps reducing the use of chemicals
- Reduction of soil compaction
- Integrated Safety
- Remote mission supervision

Size: Length: 2.69 m, track width: 1.60 to 2.10 m
Weight: 1 t
Productivity: 4 km/h, 4 ha/day.
Price: starting at €95,000

Embedded technologies:
- GPS RTK
- Electric motors
- Multicrop camera guidance for the tools

Energy: Electric Batteries

Next phase: Electric in row tools. Crop Data Collection.

Additional Information: Dino has been deployed in farms in Europe and in the USA. It is used for mechanical weeding of vegetable crops. The robot navigates the field autonomously with a great accuracy (<5 cm) and locates the crops underneath. It precisely adjusts its tools to remove the weeds as close to the plants as possible. It works on many different types of crops like lettuce, cauliflower, celery, parsley, celery, cilantro, leeks, onions, fennel. If you're willing to try it, contact us, and we will organize a demo.

ERMMI™, by Easton Robotics

About Easton Robotics

Country: USA
Date of creation: 2020
Number of collaborators: n/a
Market target: Small farms
Level of development: Under development

Core business: Agricultural robotics
Main partners: n/a
Contact: Jeff Chandler, CEO
Website: www.eastonrobotics.com

About ERMMI™

Number of robots in service: Ermmi™ is designed as a family of autonomous ground vehicles, starting with the smallest version. The company plans to develop two larger versions over time.

Main functionality: Flexible fully autonomous ground vehicles for conducting tasks common to smaller farms, including trimming, gathering, vine tending, and limited harvesting.

Features: Adjustable wheelbase, adjustable ground clearance, articulated arms with attachments for implements to carry out the specific tasks

Size: 1 m length (without arms) – smallest version

Weight: 90 kg – smallest version
Productivity: n/a
Price: n/a
Embedded technologies: GPS/GNSS/RTK, LIDAR, ROS
Energy: Battery pack
Next phase: Continued development & testing
Additional Information: The company is currently seeking funding for full development and commercialization. We expect to move to the next phase of development in 2021 and begin field testing at that time.

FAR, by Tevel Aerobotics Technologies

About Tevel Aerobotics Technologies

Country: Israel
Date of creation: August 2017
Number of collaborators: 38
Market target: Orchards
Level of development: Completion of Testing phase and Marketed

Core business: Fruit picking services
Main partners: Yaniv Maor, Eyal Desheh, Maverick ventures, OurCrowd, Forbon, Kubota
Contact: info@tevel-tech.com
Website: www.tevel-tech.com

About FAR

Number of robots in service: 20
Main functionality: Fruit picking, apples, peaches, nectarines, plumes
Features: Gentel picking, picks over 90% of fruit from tree, treetop picking
Size: 0.45 m x 0.45 m
Weight: Less than 4 kg
Productivity: n/a

Price: Service price will be equal or lower than current fruit picking costs.
Embedded technologies: AI, data fusion, perception, aeromechanics
Energy: Self powered generator
Next phase: Citrus picking, thinning, pruning, leaf cutting, selective spraying

FD20, by FarmDroid

About FarmDroid

Country: Denmark
Date of creation: 2018
Number of collaborators: 25
Market target: Organic and conventional farmers
Level of development: Marketed

Core business: Producing and developing fully automatic robots and tools to help farmers around the world.
Main partners: n/a
Contact: Eddie Pedersen, +45 28551909, epe@farmdroid.dk
Website: www.farmdroid.dk

About FarmDroid FD20

Number of robots in service: More than 150 units
Main functionality: Seeding and weeding
Features: High precision RTK-GPS, seeding system, weeding system for inter-row and intra-row weeding, CO2-neutral operation
Size: Approximately 3.8 m x 3 m x 1.5 m
Weight: 900 kg
Productivity: Up to 6.5 ha/day
Price: n/a
Embedded technologies: RTK-GPS, Siemens HMI, Solar power
Energy: Solar panels charging a lithium battery pack

Next phase: n/a
Additional Information: FarmDroid is the world's first fully automatic robot that can take care of both sowing and mechanical weed control. We help farmers and growers reduce costs of sowing and weeding crops while also doing it in a CO_2 neutral and ecological way.

FarmDroid FD20 is a solar panel driven field robot that, by using a GPS signal, marks the placement of the crops at sowing and subsequently performs mechanical weed control both between and in the rows.

La Chèvre, by Nexus Robotics Inc

About Nexus Robotics Inc

Country: Canada
Date of creation: May 2017
Number of collaborators: 11
Market target: Vegetable farmers
Level of development: Testing phase

Core business: Robot-as-a-service for weeding vegetable crops. Nexus aims to address cost and availability challenges of the labour market.
Main partner: Inertia Engineering
Contact: admin@nexusrobotics.ca
Website: www.nexusrobotics.ca

About La Chèvre

Number of robots in service: 3
Main functionality: Removing weeds near vegetable crops without damaging the crop
Features: Weeding and data collection
Size: 15 m³
Weight: 1,100 kg

Productivity: 95% weed removal
Price: Not yet determined
Embedded technologies: Codian delta robots
Energy: Hybrid
Next phase: Commercialization in Eastern Canada and Southern United States

LELY Exos, by Lely International NV

About Lely

Country: Netherlands
Date of creation: 1948
Number of collaborators: 1,750 worldwide
Market target: Grass based dairy farms
Level of development: Under development

Core business: Development and sales of auto-mated solutions for the dairy farm
Main partner: Own development
Contact: communications@lely.com (Timo Joosten)
Website: www.lely.com

About LELY Exos

Number of robots in service: 6
Main functionality: Autonomous grass harvesting and fertilization
Features: Unmanned self-propelled grass harvester
Size: 6 m x 2.5 m x 3.4 m (L x W x H)
Weight: 3,500 kg
Productivity: 5,000 – 8,000 kg fresh gras per day
Price: n/a
Embedded technologies: Stereo Vision, 4x1 Individual steering and drive, Ultrasound, GPS

Energy: Full Electric (Lithium battery packs)
Next phase: Finalize development, design freeze, first verifications in the field
Additional Information: Not only the Lely Exos atomizes the feeding of fresh grass, it also lengthens the harvesting period and can help reduce the need for roughage feed stocks. As the machine can harvest during the complete growing season of the grass, which in certain areas in the world is year-round, it enables to feed your live-stock on demand.

Phenomobile, by INRAE

About INRAE

Country: France
Date of creation: 1946
Number of collaborators: 11,500
Market target: Agriculture and Innovation
Level of development: Testing phase-Research-Experimentation

Core business: Agriculture Food & Environment
Main partners: 450 socio-economic stakeholders, businesses, associations and regional authorities
Contact: com-toulouse@inrae.fr
Website: www.inrae.fr/en

About Phenomobile

Number of robots in service: 3

Main functionality: Phenotyping robot: This robot can carry out the automated characterisation of the traits of plants found in microparcels: about crop plants, such as leaf surface area, the percentage of green leaves, leaf chlorophyll content, plant height, and the position and orientation of plant organs. These data are rendered usable thanks to automated sequential analyses.

Features: The robot is equipped with a 12-m telescopic arm fitted with an adjustable measurement probe. The probe has four synchronised cameras with five flashes, which means images can be acquired regardless of natural light conditions, yielding consistent data quality. The apparatus is also equipped with three scanning lasers (LIDAR).

Size: n/a

Weight: 8 t

Productivity: The robot is measuring 150 to 200 microparcels per hour and producing 200 MB of data per microparcel

Price: n/a

Embedded technologies: Its movements and those of its arm are controlled by RTK GPS. Measurement regimes are automatically implemented at certain points along the trajectory of the phenomobile and its arm.

Energy: n/a

Next phase: Phenomobile is currently being tested in a small field trial focused on sunflower. This work will also allow data analysis strategies for this species to be improved before the robot is deployed in a wheat-focused area that will involve 1,000 parcels.

Plantalyzer, by HortiKey

About HortiKey

Country: The Netherlands
Date of creation: 14-12-2016
Number of collaborators: 2
Market target: Professional greenhouse growers
Level of development: Marketed

Core business: Development and commercialization of data driven systems for greenhouses
Main partners: Berg Hortimotive, Royal Brinkman, Letsgrow.com, Wageningen UR
Contact: André Valstar
Website: www.hortikey.com/

About Plantalyzer®

Number of robots in service: 5
Main functionality: The most accurate crop estimation, by means of intelligent logistics, vision and AI software.
Features: The Plantalyzer® is a unique combination of autonomous robot, vision which counts and classifies tomatoes on the plant, and intelligent forecasting software. The result is reliable information for the most accurate crop estimate. The grower receives e.g. objective information about the quantity and ripeness status of fruit. It is developed in close collaboration with Wageningen University & Research and is powered by Berg Hortimotive and Royal Brinkman.

Size: 1.95 m x 0.72 m x 2.21 m
Weight: 480 kg
Productivity: 90% accurate yield prediction
Price: Available on request
Embedded technologies: Grid map, vision, data analysis, machine learning, autonomous navigation
Energy: Battery powered
Next phase: Detection of infections and diseases.
Additional Information: The Plantalyzer is a complete system that will help the grower to optimize the process between supply and demand. Thereby reducing food waste in the chain.

RoamIO Series, by Korechi Innovations Inc.

About Korechi

Country: Canada
Date of creation: 2016
Number of collaborators: 8
Market target: Field Crops, Orchards, Vineyards, Vegetable crops
Level of development: Marketed

Core business: Robots to automate tasks in outdoor farms, green houses and vertical farms
Main partners: Haggerty Creek Inc., SoilOptix Inc., Ontario Centre of Innovation, Durham College, Niagara College
Contact: Sales@Korechi.com | +1 289-700-3997
Website: www.korechi.com

About RoamIO Series

Number of robots in service: 3
Main functionality: Automates data logging for soil measurements, yield estimation and pest & disease scouting; implements such as aerators, seed & fertilizer spreaders, mowers, soil samplers and seed planters; harvest assistance
Features: Multi-purpose tracked robotic platform, highly adaptable, field-programmable through a tablet computer with an intuitive user-interface. RoamIO is equipped with a state of the art path generator and has multi-layer collision avoidance. RoamIO is already enabled with multi-band GNSS - RTK waypoint navigation with up to 5 cm repeatable navigation precision. Soil compaction under 3 PSI.
Size: Customizable, RoamIO-HCT: 1.6m x 1.7m x 1.4m (L x W x H)

Weight: Variable, 550kg for RoamIO-HCT
Productivity: Up to 15 h/day in Vineyards and 40 h/day in field crops
Price: Starts at USD 40,000
Embedded technologies: LiDAR, Machine Learning, Computer Vision, Waypoint navigation, RAPG
Energy: 14.4 kwhh Lithium Battery–optimal performance for up to 3,000 charge cycles, optional petrol/ diesel generator to extend range
Next phase: Carry out more pilots for the platform and implements
Additional Information: Looking for distributors in USA and Europe

Robot One, by Pixelfarming Robotics

About Pixelfarming Robotics

Country: The Netherlands
Date of creation: 2019
Number of collaborators: 3
Market target: Open cultivation
Level of development: Testing phase
Core business: We design and manufacture advanced agricultural robots to support biodiverse farming.

Main partners: Provence Noord Brabant, Idea-X and Campus Almkerk
Contact:
Pixelfarming Robotics, Laagt 16, 4286 LV ALMKERK
info@pixelfarmingrobotics.com
Tel: +31852733075
Website: www.pixelfarmingrobotics.com

About Robot One

Number of robots in service: 3 in the season of 2021
Main functionality: Pixelfarming Robotics' Robot One is a smart agricultural robot for biodiverse farming, designed to control weeds without the use of pesticides. It is designed to control plants and weeds autonomously, based on Vision and ready to be equipped with tools for specific crop treatment.
Features: Autonomous, all electric, lightweight and accuracy of 2cm
Size: 3.5 m x 1.7 m x 2.3 m (L x W x H)
Weight: Approximately 1,100 kg
Productivity: Approximately 1 ha/h
Price: €150,000 - €185,000

Embedded technologies: Dual GPS RTK, 2 cm accuracy CPU Nvidia Jetson Xavier, 8 core 64 bit CPU, 512 core GPU, 16 GB 256 bit memory Connectivity 4G, 100 Mbps Camera 4 x Intel D415
Energy:
- Drive system:
  - Voltage 48VDC Torque 500 Nm per wheel
  - Maximum speed 1 m/s Position accuracy
- Robot system:
  - Performance peak power 1900 W, 300 rpm
Next phase:
- Validation of field tests in several types of crops
- Extending the tools for the upgradable platform

Robotti 150D, by Agrointelli

About Agrointelli

Country: Denmark
Date of creation: 2015
Number of collaborators: n/a
Market target: Farmers growing row crops

Level of development: Marketed
Core business: Robot manufacture
Contact: Frederik Rom, Sales Manager at FRR@agrointelli.com or at +45 9189 1181
Website: www.agrointelli.com

About Robotti

Number of robots in service: 20
Main functionality: Seedbed cultivation, seeding, mechanical weeding and band-spraying / spraying
Features: Reliable and powerful diesel and hydraulic system, with a standard 3 point hitch, hydraulics, PTO and RTK GPS.
Size: Can be customized towards user needs: working width between 1.5-3.5 meters
Weight: Approximately 3,000 kg

Productivity: Operates successfully between 0-8 km/h
Price: Between €160,000 and €185,000 (depends on extra equipment)
Embedded technologies: Redundant safety system including Lidar scanners, emergency stops and cameras for live monitoring. Also, a smart route planning system is provided that optimize the working route in the field.
Energy: 150 Hp.
Next phase: n/a

Romi Rover, by Romi Organisation

About Romi Organisation

Country: France
Date of creation: 2020
Number of collaborators: 7
Market target: Organic market farms
Level of development: Testing phase
Core business: Open robotics platform for organic market farmers

Main partners: Sony Computer Science Laboratories, Chatelain Maraîchage, France Europe Innovation, Institute of Advanced Architecture of Catalonia (Iaac) / FabLab Barcelona, CNRS, Inria, Humboldt University of Berlin
Contact: info@romi-project.eu
Website: www.romi-project.eu

About Romi Rover

Number of robots in service: 3
Main functionality: Mechanical weeding
Features: n/a
Size: 1.42 m x 1.36 m x 1.45 m (W x D x H)
Weight: 120 kg
Productivity: 600 m²/day (precision weeding inter-row & intra-row), 7,200 m²/day (classical weeding, inter-row)

Price: Between €5,000 and €10,000
Embedded technologies: HW: Raspberry Pi, Arduino, 8MP RGB camera. SW: Support Vector Machine for image segmentation, custom software.
Energy: Lithium batteries
Next phase: Beta-testing & certification

Spoutnic, by TIBOT Technologies

About TIBOT Technologies

Country: France
Date of creation: 2016
Number of collaborators: 12
Market target: Growers & large Integrators
Level of development: Marketed
Core business: Poultry robotics

Main partners: Demeter, Seventure Partners, Breizh-Up, Crédit Agricole Ille-et-Vilaine Expansion, Caisse Régionale du Crédit agricole d'Ille-et-Vilaine, BNP Paribas and le Crédit Mutuel de Bretagne
Contact: contact@tibot.fr
Website: www.tibot.fr

About Spoutnic

Number of robots in service: Around 200 units
Main functionality:

- It decreases strenuous activities for the poultry growers.
- It moves the birds, reduces the floor-egg ratio.
- It improves fertility (+5 points fertility and +30% in matting activity).
- It improves the profitability of operation.

Features:

- Spoutnic is plug & play.
- Starts with two buttons.
- Do not require any technical installation.
- Adapts to any operation: Turkeys, broiler breeders or layers thanks to its customization controller!

Size: 0.63 m x 0.58 m x 0.19 m
Weight: 12 kg

Productivity: 10 to 12 hours autonomy
Price: €6,800
Embedded technologies:

- Adaptable speed and maneuver angle.
- Wide range of sounds and lights to stimulate the birds.
- Bump and turn navigation system.

Energy: Lithium battery

Next phase: Thanks to our continuous customer feed-back process, we are now able to prepare for multiple evolutions of our product. The next TIBOT Technologies robots are coming up.

Additional Information: Spoutnic's compact size allows it to go under the feeding lines. If it encounters an obstacle, it detects it and performs a bypass maneuver. Its 6 speeds and programmable maneuver angle allow it to adapt to any batches.

SwarmBot, by SwarmFarm Robotics

About SwarmFarm Robotics

Country: Australia
Date of creation: 2014
Number of collaborators: n/a
Market target: All of Agriculture
Level of development: Marketed

Core business: Agricultural Robotics
Main partners: n/a
Contact: Andrew Bate
Website: www.swarmfarm.com

About SwarmBot

Number of robots in service: 14
Main functionality: Spraying and mowing
Features: 75hp 4 wheel drive
Size: Length 4.0 m x Height 1.9 to 2.4 m depending on configuration
Weight: 2,500 kg

Productivity: 10 km/h
Price: AU$70,000/year for 3 years
Embedded technologies: n/a
Energy: Diesel
Next phase: n/a

The Digital Farmhand, by AGERRIS

About AGERRIS

Country: Australia
Date of creation: 2019
Number of collaborators: n/a
Market target: Vegetable farmers
Level of development: Marketed

Core business: Robotic Smart Weeding and Crop Intelligence
Main partners: n/a
Contact: enquiries@agerris.com
Website: www.agerris.com

About The Digital Farmhand

Number of robots in service: 10

Main functionality: The Digital Farmhand uses the on-board crop intelligence system to collect information at an individual plant level from the first growth stage.

It will learn about the different crops that have been planted and the factors that impact crop health and yield, and report back on the location and volume of crops versus weeds.

Once the Digital Farmhand has been trained, it can act in real time using intelligent precision tools.

Today's robot will chip out weeds as it travels along the vegetable beds leaving the crops to grow in a healthy and chemical free environment.

Features: The Digital Farmhand is solar-electric and can run in autonomous mode. It has a battery life of up to 15 hours with solar panels, can carry up to 200 kg, at speeds up to 6 km/h. Safety features include obstacle detection and remote/onboard emergency stop.

Size: Adjustable from 0.9 m (3 ft) to 1.8 m (6 ft), or multi-span for wider options.

Weight: Up to 400 kg (depending on tools)

Productivity: Up to 3 ha/day

Price: On application

Embedded technologies: Fully autonomous navigation options using dual frequency RTK GPS, stereo vision and laser. Crop detection and machine learning algorithms control real time intelligent weeding tools.

Energy: Electric powered – 360W solar; 2x24V Lithium Iron Phosphate batteries @ 100 Ah per battery

Next phase: Working with tree crops.

Titan FT-35, by FarmWise

About FarmWise

Country: USA
Date of creation: May 2016
Number of collaborators: 50
Market target: Vegetables
Level of development: Marketed

Core business: Automated mechanical weeding
Main partners: A dozen large-scale vegetable farms from the Central Coast of California and Arizona
Contact: contact@farmwise.io
Website: www.farmwiselabs.com

About Titan FT-35

Number of robots in service: n/a
Main functionality: Intercrop and inter-row mechanical weeding
Features: Cultivation tools
Size: n/a
Weight: n/a

Productivity: 10-15 acres a day
Price: n/a
Embedded technologies: n/a
Energy: n/a
Next phase: Data platform for farmers, optimization of other farming tasks

Tom, by Small Robot Company

About Small Robot Company

Country: UK
Date of creation: 2017
Number of collaborators: 50 employees
Market target: Arable, wheat in the first instance and later all arable crops including barley, oats, soya, maize and rice
Level of development:
- Under development (Dick and Harry robots; SlugBot)
- Testing phase (Tom robot)
- Marketed (Wilma robotics Advice Engine)

Core business: Agri-robotics
Main partners: Rootwave (non-chemical weeding); Crop Health and Protection Centre (SlugBot)
Contact: info@smallrobotcompany.com
Website: www.smallrobotcompany.com

About Tom

Number of robots in service: 5
Main functionality: Crop monitoring for weeds, disease, pests, yield prediction and fertiliser requirements
Features:
- 5.5 m boom
- 6 m camera coverage
- 6 x 5K cameras
- RGB and hyperspectral image capture
- 1.5 m/s forward speed
- 4 wheel drive
- Static ground pressure 33.7 kpa

Size: 1.315 m x 1.350 m x 1.909 m (H x W x L)
Weight: 232 kg
Productivity: 20 ha/day

Price: TBC
Embedded technologies:
- Nvidia Jetson
- 3 x 1.56 kwh rechargeable batteries
- Swiftnav GPS system
- 2 x GNSS receivers

Energy: 3 x 1.56 kwh batteries with battery life 4 hours
Next phase: Commercial deployment to core farmer advisor customers autumn 2021
Additional Information: We are not going into large-scale manufacture and commercial deployment until 2023 when R&D for all three of our robots - Tom, Dick and Harry is complete.

TREKTOR, by SITIA

About SITIA

Country: France
Date of creation: 1986
Number of collaborators: 25
Market target: TREKTOR for vineyards, market gardening, tree crops
Level of development: Marketed

Core business: Agricultural robotics, & EC test rigs for automotive and aeronautic industries
Main partners: French Agriculture chamber, Research university (INRAE), Administry of agriculture, Robagri, AXEMA
Contact: agri@sitia.fr
Website: www.sitia.fr/en/innovation-2/trektor/

About TREKTOR

Number of robots in service: 10 in 2020
Main functionality: Autonomous work with mechanic, hydraulic or electric implements in specialized culture (vineyard, market gardening, orchard...).
Features: Hybrid electric/diesel, 24h/autonomy, variable track and height to adapt to different crops, up to 35L/min at 180bars hydraulic power, obstacle detection, autonomous navigation

Size: 3 models: Mini, Midi or Maxi, overall width from 1.39 m to 2.75 m depending on the model
Weight: Up to 3 t depending on the model
Productivity: 24/7
Price: To be asked at your local distributor or agri@sitia.fr
Embedded technologies: n/a
Energy: Hybrid: Electrical actuators & Diesel engine for battery charging
Next phase: Worldwilde distribution

VineScout, by Agricultural Robotics Laboratory

About Agricultural Robotics Laboratory

Country: Spain

Date of creation: 2012

Number of collaborators: 5

Market target: Equipment manufacturers, digital agriculture service providers

Level of development: Testing phase

Core business: Research, consulting, and IP licensing (Public University)

Main partners: Varta, Disai, Intecdes

Contact: Francisco Rovira Más: frovira@upv.es

Website: www.agriculturalroboticslab.upv.es

About VineScout

Number of robots in service: 1

Main functionality: High-resolution vineyard monitoring

Features: VineScout is a ground robot for vineyard monitoring and management. The robot moves autonomously inside vineyards structured in vertical trellises and produces maps of relevant production parameters such as water status and vine vigor. The data are processed by a model that generates maps for vine growers and wine producers to make decisions on irrigation or harvesting.

Size: 1.2 m x 1 m x 1 m (L x W x H)

Weight: 200 kg

Productivity: 5 ha/day

Price: Not in production yet

Embedded technologies: The robot is designed to monitor one canopy side every two rows. Autonomous navigation is achieved by the fusion of real-time data fed by a time-of-flight 3D sensor, a lidar, and four ultrasonic sensors. These perception sensors also assist in safeguarding tasks. The robot carries an ambient sensor for measuring air temperature, pressure, and relative humidity, and three crop sensors for measuring NDVI, PRI, and canopy temperature.

Energy: The robot is powered by three Li-ion batteries providing 192 Ah.

Next phase: Wide validation of crop assessment models. Refining of navigation & handling capabilities.

Additional Information: www.vinescout.eu/web

Weed Whacker robot, by Odd.Bot

About Odd.Bot

Country: The Netherlands
Date of creation: 2018
Number of collaborators: 21
Market target: (Organic) Open Cultivation Vegetables
Level of development: Testing phase

Core business: Weed detection & removal
Main partners: Wageningen University & Research (WUR), TU Delft, BioNext
Contact: info+FIRA2020book@odd.bot
Website: www.odd.bot

About Whacker robot

Number of robots in service: 2
Main functionality: Autonomous Mechanical in-row Weeding for high-density crops
Features: Stamper/Pusher, Puller, Plucker
Size: 4.3 m x 1.5 m x 2.1 m approx. (L x W x H)
Weight: 1,500 kg
Productivity: n/a
Price: €1,200 per ha per crop cycle / 2
Embedded technologies: A.I./Machine Learning & Delta arm robots

Energy: Electric
Next phase: Sign up for the Trailblazer program, the first pilot program for 2021
Additional Information: Smart & Sustainable Weeding! Exact Plant Specific. The Weed Whacker robot provides a higher yield with less manual labour. The solution contributes to a world with less chemical herbicides and in particular glyphosate. For a flourishing future. Autonomous Mechanical In-Row Weeding by Odd.Bot

FIRA 2020
Thank you to our main partners

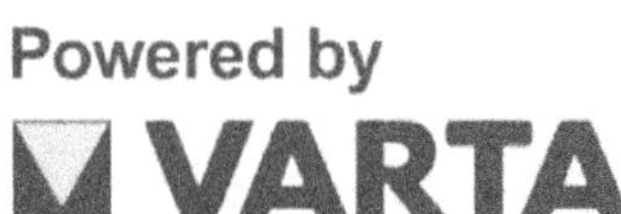

Layout: Hélène Bonnet Studio 9
Imprimé pour vous par Books on Demand (Allemagne)